PRACTICAL HYDRAULICS HANDBOOK

Barbara A. Hauser

 LEWIS PUBLISHERS

Library of Congress Cataloging-in-Publication Data

Hauser, Barbara A.
  Practical hydraulics handbook / Barbara A. Hauser.
    p. cm.
  Includes bibliographical references.
  1. Water-pipes--Hydrodynamics. 2. Sewerage--Fluid dynamics.
  3. Waterworks. I. Title.
  TC174.H38 1991
  828.1'5--dc20   91-10248
  ISBN 0-87371-548-9

LEWIS PUBLISHERS, INC.
121 South Main Street, Chelsea, Michigan 48118

PRINTED IN THE UNITED STATES OF AMERICA

ACKNOWLEDGEMENTS

My thanks to Norm Clark, Bob Richards, and Mike Nantelle, Computer Aided Design Department at Bay de Noc Community College, for their many hours of assistance with graphic design, and to Jim Lundberg, Computer Science department, for processing assistance.

I am grateful to Warren Isaacson, City of Escanaba, and John Erickson, Michigan Department of Public Health, for help with technical details, and to Otto Green, City of Bay City, and Dr. Ed Cory, Ferris State University, for their encouragement and guidance during the preparation of this work.

Interior design by Jennifer Gardner.

ABOUT THE AUTHOR

With a Baccalaureate Degree in Biology and Chemistry from Trinity College in Washington D.C. in 1962, Barbara Hauser began her career as a biochemist at the Rockefeller Institute in New York City, and then at the University of Maine, at Orono. She then "retired" to raise a family of four children, until a move to Upper Michigan reactivated her career. Considering an education update first priority, she entered Bay de Noc Community College in Escanaba as a student, and earned an Associate of Applied Science degree in Water Purification Technology. During the following years Mrs. Hauser was employed by the City of Gladstone, MI, in municipal wastewater treatment, and by the Rexnord Company, Research & Development Group, Milwaukee, WI, as an operations specialist in advanced wastewater treatment. She has also earned a Master of Science degree in Occupational Education from Ferris State University, Big Rapids, MI.

For the past eight years, Mrs. Hauser has been employed as department head and chief instructor of the Water Purification Technology department at Bay de Noc Community College. She emphatically states, "Summer is not the time when instructors sleep!", and her summer months have been spent running specialized training courses for industrial wastewater treatment plant operators, organizing regional seminars in Biology and Chemistry for municipal water utility operations personnel, teaching science and computer courses at the college, and working on research projects with the local Department of Natural Resources.

Mrs. Hauser received the "Outstanding Occupational Educator" award, sponsored by the Michigan Occupational Dean's Council, in 1990. She has previously been published in *Public Works* magazine.

TABLE OF CONTENTS

INTRODUCTION

Controlling the flow of water from one place to another, the proper amount at the proper pressure and velocity, is perhaps more taken for granted than any other aspect of science. We marvel at the colors and dramatic effects produced in a chemistry magic show. We are awed at the bloom of a flower, but when we awake in the morning to shower and shave, do we ever wonder if water will come out of the faucet today?

Potable water systems are taken for granted, but they must be carefully designed and controlled by the water utility in order to provide adequate pressure, safety from backflow contamination, and proper water velocity for protection of household piping.

Water is a carrier for most household and industrial wastes. Proper hydraulic design and maintenance of wastewater collection systems assures sanitary transport to the treatment plant and prevention of sewer backup.

A new operator at a water or wastewater utility is most concerned with learning the daily routine of laboratory testing, operation and process maintenance. The hydraulics of his treatment process has been designed into it. This is what keeps the water moving from one process unit to another, providing correct detention times, proper settling velocity, lift to a higher elevation, etc. His consideration of hydraulic principles may at first be minimal - until the electrical power goes out, a pump malfunctions, a line breaks, or a blockage occurs. When hydraulic control is no longer there, treatment process grinds to a halt. No water utility can operate without it.

In as simple a manner as possible, this text will cover the principles and calculations dealing with the hydraulics of water systems. Since the subject area can become quite complex, we will stress only what is necessary for a basic understanding, and we will emphasize practical applications for water and wastewater utility operations. We will deal essentially with the flow of water, and only occasionally with the control of other liquids or gases.

Advanced mathematics will not be attempted; calculations will be limited to arithmetic conversions, basic algebra, exponential notation. Some previous knowledge in these areas will be assumed, as will knowledge of a few formulas: area of a circle and rectangle, volume of containers of these shapes. It will be assumed that a scientific calculator is being used.

The first chapter discusses some basic and frequently encountered properties of fluids as we experience them around us. It does not include calculations, and is meant to bring to mind recognized phenomena, and provide a general understanding of fluid energy and movement. References to these concepts will appear throughout the text and will be elaborated upon where appropriate. More specialized terminology will be introduced with each chapter, as applicable.

Succeeding chapters are presented in workbook style, with emphasis on solutions to practical problems. However, I do remember that the only math book that I have ever enjoyed reading was one which had absolutely no math in it. The principles were explained well enough to provide a solid grasp of the concept, I was not forced to work with any numbers, and I actually wanted to go on to the next chapter. I have set up this text in a similar fashion. It could be read without ever doing any of the problems. The body of each chapter holds an explanation of the principles to be covered. The applied mathematical section is at the end, and includes a list of problems derived from the concepts, with answers, and with a detailed solution to each problem.

I have attempted to set up each solution in as readable a form as possible. In doing so, some shortened forms are in use. The reader should be aware of these:

--Most values are converted to units of "feet" for calculation. (This includes - feet, square feet, feet per second, cubic feet per second.)
--Unit labels are often omitted in the body of a solution, but are included with the final answer.
--Multiplication signs are often omitted in the body of a solution involving a formula.
--Most answers are rounded off to one or two decimal places.
--In diagrams, the pressure gage symbol ♀ with a number inside it designates a pressure reading in psi (pounds per square inch).

PRACTICAL HYDRAULICS HANDBOOK

CHAPTER 1

THE BASICS

Matter is divided into two major classes: Solids and Fluids.

Solid - matter which holds its shape and will not deform without external stress.

Fluid - a substance that will deform without external stress, and takes the shape of its container. Both liquids and gases are fluids. A fluid is infinitely flexible; it can move rapidly in one place and slowly in another. It can transmit a force in any or all directions. No other medium has such ability to transmit maximum power with a minimum of bulk and weight.

Hydrology - the study of the naturally occurring waters of the earth. It includes their circulation, distribution, and reaction with their environment.

Hydraulics - the study of the principles that govern the behavior of liquids at rest and in motion. It is the study of the mechanics of water and its control by man.

Pressure - an amount of force acting on a unit area. Pressure is the force that moves water in a closed pipe system, and is often registered as lb./sq.in. (psi), or lb./sq.ft. (psf). Pressure at a given point registers with equal intensity in all directions, and is measured with various forms of gages, which may represent pressures above or below atmospheric pressure. Pressure can also be registered in terms of feet of water, for its origin is the weight of a depth of water directly resting upon the area of measurement. When pressure is measured in feet, it is called Head.

Friction - the molecules of a fluid not only have an attraction for each other, but also for the molecules of any solid which they come in contact with. The attraction causes the speed of the fluid molecules to approach the speed of the solid molecules, as they collide. Because they are not as rigidly connected to each other as the solid molecules are, the fluid molecules become disorganized, moving at different speeds and in different directions, resulting in the turbulence which is friction. Because of this, energy is decreased; pressure is lost.

Head Loss - a loss of pressure, registered in feet of water. It is the energy loss encountered in a moving water system which occurs because of friction created as the water rubs against the sides of the conduit and passes through fittings.

Compressibility - a reduction in volume of fluid which occurs with the application of external force, such as pressure. Gases tend to fill the entire container which they are in; they are easily compressible; their density varies with pressure. Most liquids are not. When held in a container, liquids fill the bottom area and form a well defined surface. Density remains constant, and the liquid will fill a specified volume. For all practical purposes, water is an incompressible fluid. With an application of 100 psi pressure, water occupies a volume of only .0000023 cubic feet less than it does at atmospheric pressure. This is very important in the design of piping systems which must handle stresses which the water cannot absorb by compression.

Viscosity - internal resistance of the fluid to the motion of its particles. A viscous fluid tends to damp out turbulence. The fluid moves in streamlines and resists the shear force that causes friction. This principle is in use with lubricants. Because they are viscous, the lubricant layers slide upon each other evenly, minimizing turbulence and the resultant creation of heat. Many liquid plastics are very viscous. Water is very nonviscous. Its viscosity is 1.41×10^{-6} sq.ft./sec.

Velocity - the speed at which a liquid moves. It may be expressed as ft./sec., ft./min., meters/sec. Note that an expression of velocity alone designates nothing about how much of the liquid is moving - only how fast it is moving.

Fluid Flow is Governed by Three Fundamental Concepts:

Conservation of Mass - material is neither created nor destroyed. Whatever enters the system must leave it, or it will accumulate inside. Since water is incompressible, it cannot accumulate in a pipe.

Conservation of Energy - there are two kinds of energy, kinetic and potential. In a water system, kinetic energy is demonstrated by the velocity of the fluid. Potential energy is present as elevation of the fluid above a reference point, or as pressure. The summation of these three possible energies in a water system is expressed in an equation known as Bernoulli's Theorem, which is the basis for all hydraulic calculations. This will be taken up throughout the text.

Conservation of Momentum - momentum is the mass of the fluid, multiplied by its velocity. Perhaps we could think of it as the amount of "oomph" the water has as it travels. Upon encountering bends, turns, obstructions in the conduit, momentum is decreased, but it is converted to a temporary increase in pressure, which is then dissipated as heat. The effect of this is thrust, surge, or water hammer, which can cause serious damage to pipes, pumps and fittings if not provided for. These three are similar phenomena and yield the same result in transient pressures created:

Thrust - force created at turns in a pipeline; it tends to push the fitting loose.

Surge - slow motion mass tendency of the water to "pile up" as momentum decreases. It results in a pressure increase.

Water Hammer - temporary great increase in pressure resulting from a rapid change in velocity or direction of the water - often heard as a loud "bang" upon closing a valve.

Fluid Flow - flow is the quantity of fluid which passes a point in a given unit of time. It is expressed most frequently as cubic feet/sec., gal./min., gal./day, or Million Gallons per Day (MGD). Flow can be viewed as a moving volume. Expressing flow quantities gives no indication as to the speed of the water, just how much of it is moving.
--Flow is complex and not always subject to exact mathematical analysis.
--Flow may be steady or unsteady.
--Flow may by uniform or nonuniform.
--Flow may be laminar or turbulent.
--Flow may occur in closed systems, under pressure, or it may be open channel flow.

Steady State Flow - flow is steady if, at any one point, the velocity of the water particles is the same and the pressure is constant. Calculations with unsteady flow conditions are complex, and will not be covered in this text.

Uniform Flow - occurs when the magnitude and direction of velocity do not change from point to point. The flow of water through long pipelines of constant diameter is uniform. Non-uniform flow occurs when velocity, depth, pressure, etc. change from point to point, such as when a pipe exits into a large tank (depth increases, velocity decreases), or when a constricted area is encountered in a pipe (velocity increases, pressure decreases). We will consider both uniform and non-uniform flow in this text.

Laminar Flow - where laminar flow occurs, fluid particles move along straight parallel paths, in layers, or streamlines. Their velocities need not be the same. In a pipe, the outside layers rub against the pipe walls, and will travel more slowly. In true laminar flow, the boundary layer does not move at all, and all inner layers slide upon it. There is no turbulence in the water, and no significant pressure loss due to friction. Laminar flow is very difficult to achieve, and when water is the fluid, only occurs at extremely low velocities. (For any fluid, there is a velocity below which the flow is laminar; it is called Critical Velocity). Viscosity and pipe diameter are also factors. Laminar flow may occur in pipes carrying viscous oils at low velocities, and in tubes of very small diameters. It is not typical of normal water pipe flow, which is turbulent. But we do see it as groundwater flow; density layers remain distinct; contaminants do not dilute. An attempt at laminar flow conditions is demonstrated in the use of tube and plate settlers for clarification. The flow is forced to hold streamlines, and solids are more likely to settle out. The components of laminar flow have been combined to create a value, called the Reynold's Number, R. As applied to fluids, it is a function of velocity, diameter, and viscosity. (R = velocity x diameter/viscosity). If R is less than 2000, flow is laminar. Reynolds Numbers in the hundreds of thousands are the norm for pipe flow in water systems.

Turbulent Flow - when flow is turbulent, the water particles move in a haphazard fashion and continually cross each other in all directions. It is impossible to trace the motion of individual particles, and pressure losses are encountered along a length of pipe because of it. Turbulent flow is normal for a water system.

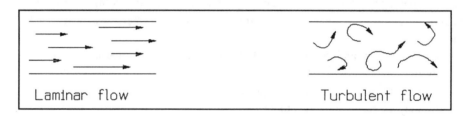

Laminar flow   Turbulent flow

Open Channel Flow - an open channel is a conduit in which the liquid flows with a free surface subjected to atmospheric pressure only. Open channel flow is caused by the downward slope of the channel, which maintains the velocity. Flow in aqueducts, sewer pipes, streams and rivers is open channel flow.

Hydraulic Jump - this is an open channel flow phenomenon. When water traveling at high velocity encounters conditions which reduce the velocity, the depth of the water surface increases, and the water surface "jumps" to the greater depth. For instance, water coming over a dam as a shallow flow at high velocity, at the bottom of the dam can hold the deeper river flow at  bay. The jump acts to dissipate the velocity of the liberated water from the dam, and then it continues downstream at the greater depth. This is usually quite a dramatic phenomenon, accompanied by turbulence. Its position must be planned into spillway design at a dam, in order to prevent river scour downstream. The turbulence and visible depth increase at the bottom of a flume is an example of hydraulic jump. So are shallow waves, such as those created at the wake of a passing boat. Shoreline waves, and flash floods, which may create an approaching "wall of water", are examples of traveling hydraulic jump. Hydraulic jump may be made useful to mix treatment chemicals, or to aerate water for purification.

Capillarity - the rise or fall of a liquid in a capillary tube, or in porous material, caused by surface tension (attraction of the water molecules to each other, and to the molecules of the tube). We note this effect in the ability of a sponge or paper towel to soak up water, the rise of kerosene into the wick of a lantern, and in groundwater, in the semisaturated volume of aquifer just above the water table.

Vapor Pressure - the partial pressure created by water vapor molecules as the liquid evaporates. It can also be viewed as the liquid's tendency to evaporate. Vapor pressure in most cold water systems is insignificant. The factor becomes important when dealing with hot liquids, at very high elevations, or when inpipe conditions approach absolute vacuum (suction lift in pumping applications).

Force - pressure registered upon an entire area, such as the force of water on the side of a dam, or the hydraulic force upon the walls and bottom of a basement which may cause it to collapse when the ground is flooded. Force is registered in pounds.

Power - in a water system, power is usually referred to as the energy supplied by a pump. It is measured in foot pounds per second. Large units of power are called Horsepower (550 ft.lb./sec.). The ability of a pump to move water is usually judged by its horsepower rating.

Flow Nets and Flow Models - these are complex patterns simulating the changes in water pressures and flow within a specified area. They are often multi-dimensional, are now most frequently computerized rather than manually created, and are beyond the scope of this text. Typical examples would be:
--The entire drainage pattern of a watershed.
--The travel and dispersion of a contaminant plume in an aquifer.
--A model of an entire water distribution system with pressures and crossflows indicated.

CHAPTER 2

MASS, DENSITY AND DISPLACEMENT

MASS & WEIGHT

Mass is the quantity of matter that a substance contains. It is a basic property of the substance, and is constant, regardless of location. Mass is most frequently registered in units of grams, or pounds; it is measured with an analytical balance which compares it against a known mass.

Weight is the effect of the force of gravity upon a substance, and is measured with a scale. A block of steel which weights ten pounds on earth, will weigh much less on the moon, where gravitational forces are less. Its mass, however, will be the same in both places.

Since we are dealing only with earthly water systems, for our purposes mass and weight are the same quantity.

DENSITY

Also called SPECIFIC WEIGHT, density is mass per unit volume, and may be registered as lb./cu.ft., lb./gal., grams/ml., grams/cu.meter. Taking a fixed volume container, filling it with a fluid, and weighing it, we can determine density of the fluid (after subtracting the weight of the container).

<u>One cubic foot of water weighs 62.4 pounds</u>. (Density = 62.4 lb./cu.ft.)

<u>One gallon of water weighs 8.34 pounds</u>. (Density = 8.34 lb./gal.)

<u>One milliliter of water weighs 1 gram</u>. (Density = 1 gram/ml.)

TEMPERATURE

The density of materials is affected by temperature. Many solids expand when hot. Fluids also become less dense when warm. Water has the peculiar property of being most dense at four degrees Celsius. Above and below that temperature, it expands. This accounts for seasonal turnover in lakes, and the formation of density gradients, or layers, in reservoirs and behind dams, often concentrating pollution at specific depths. In settling tanks and clarifiers, this effect can inhibit uniform mixing of influent water, causing short circuiting. Most solids will settle more effectively in warm waters than in cold waters because the water is denser when cold.

PRESSURE

Though the density of gases is greatly affected by pressure, the density of most liquids, including water, is not. For all practical purposes, water is considered an incompressible fluid; 62.4 pounds of water occupies a volume of one cubic foot, regardless of the pressure applied to it.

SPECIFIC GRAVITY

This value designates the ratio of the density of a substance, compared to the density of a standard substance. It is a way to specify relative densities. For liquids and solids, the standard chosen is the weight of water. Therefore:

$$\text{Specific Gravity} = \frac{\text{Density of Substance}}{\text{Density of Water}}$$

If the Specific Gravity of water were to be determined this way, we would write:

$$\text{Specific Gravity} = \frac{\text{Density of Water}}{\text{Density of Water}} = 1$$

The Specific Gravity of water is 1.

Substances with a Specific Gravity of less than 1 will float (gasoline, styrofoam, wood, wastewater scum). Substances with a Specific Gravity of more than 1 will sink (bricks, steel, grit, floc, sludge). A good example of treatment process taking advantage of different Specific Gravities is the multi-media filter. During backwash, the different grades of media are mixed up, but at the end of the backwash cycle, as the water quiets, the media layers become separate and distinct: anthracite on top, sand in the middle, garnet on the bottom. These substances are chosen for filtration partially because of their different specific gravities.

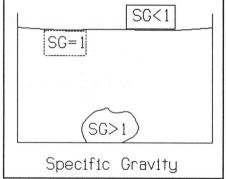

Specific Gravity

An anaerobic digester is another good example of the effect of Specific Gravity. Separation of solids, liquid and gas is essential to operation, enabling each to be drawn off separately.

Just as the density of liquids and solids are compared to water as the standard, the Specific Gravity of gases is based on the density of air. Air weighs .075 pounds/cubic foot at standard temperature and pressure. Specific Gravity calculations are done in the same manner, and will determine whether a gas rises or falls in a room full of air.

$$\text{Specific Gravity} = \frac{\text{Density of Gas}}{\text{Density of Air}}$$

Once Specific Gravity is known, density and weight can be calculated. Note that Specific Gravity is a pure number with no unit label. The units have cancelled out in the calculation.

DISPLACEMENT

A solid submerged in a body of water will displace its own volume. The displacement causes a rise in water level equivalent to the volume of the submerged object. An object partially submerged displaces a volume of water equivalent to the submerged section of the object.

Sometimes it is useful to know the volume of an irregularly shaped object. This can be determined by measuring the displacement. For example, to calculate the percent volume of mudballs in a sand filter, take a core sample with a small cylinder of known volume. Extracting a portion of media, sift out the sand, leaving only the mudballs. Pour them into a large graduated cylinder, filled to a liter with water. A volume of water will be displaced in the cylinder equivalent to the volume of the mudballs. Record the milliliter rise in the cylinder, divide by the volume of your core sampler, and you have your percent mudballs - in your sampler, and in your sand filter.

BUOYANCY

Buoyancy is known as Archimedes' Principle, and has been employed by man for over two thousand years. An object submerged in a liquid is subject to an upward force equal to the weight of the volume of liquid it has displaced. A dramatic example of this is the underwater swimmer, whose body is lighter than the weight of the water that he is displacing. He actually assumes a negative weight underwater, which causes him to fall upward, unless he purposely swims down. This force of buoyancy applies to any submerged object. For example, a cubic foot of steel weighs 486.7 pounds in air. But immersed in water it weighs 486.7 lb. - minus the weight of the cubic foot of water it has displaced (the force of buoyancy).

PROBLEMS

1. The water in a tank weighs 820 pounds. How many gallons does it hold?

2. If your municipal water rate is one dollar per thousand gallons, how many pounds of water do you have delivered to your house - for a dollar?

3. Three cubic feet of gasoline weighs 131 lb.
   A. What is the Specific Gravity of gasoline?
   B. How much does a gallon of it weigh?

4. If an oil weighs 55 lb./cu.ft., how much does a five gallon can of it weigh?

5. You have a truck designed to transport 5000 gallons of liquid.
   A. How many pounds of water can you transport?
   B. How many pounds of sulfuric acid (SG=1.83)?

6. What is the Specific Gravity of concrete if it weighs 150 lb./cu.ft.?

7. The Specific Gravity of mercury is 13.6.
   A. What is its weight per cubic foot?
   B. Per cubic yard?

8. If the density of a sand is 100 lb/cu.ft., how much does a cubic yard of this sand weigh?

9. Chlorine gas is 2.5 times heavier than air.
   A. What is the weight of a cubic foot of chlorine?
   B. What would a room (10' x 12' x 8') full of chlorine gas weigh?

10. You are pumping a solution of alum (density = 83 lb./cu.ft.) at a rate of 3 gpm into a chemical treatment system.
    A. How many pounds are fed in one day?
    B. What is the Specific Gravity of the alum solution?

11. A digested sludge whose specific gravity is 1.25 is pumped at 50 gpm to the drying beds.
    A. How many pounds of digested sludge are applied to a bed which is 30 ft. wide, 60 ft. long, and 10 inches deep?
    B. How long does it take to fill this bed?

12. A solid piece of plastic whose specific gravity is 1.2 is dropped from a boat into a lake. How deep will it sink?

13. A stone weighs 90 lb. in air. When immersed in water, it weighs 50 lb.
    A. What is the volume of the stone?
    B. What is its Specific Gravity?

14. If a 150 lb. person swims in a pool, how many cubic feet of water will he displace if his Specific Gravity is 1.1?

15. A heat exchanger is being constructed to maintain temperature of the anaerobic digester. Sludge will pass through a coiled 4 inch diameter pipe 160 ft. long, which is set inside a hot water tank 4 ft. by 4 ft. by 4 ft. high. How many gallons of water will be needed to fill this tank?

16. An open box is to be sunk to its rim in water. If its dimensions are 10 ft. by 10 ft. by 8 ft. deep, how many pounds must it weigh in order to stay submerged?

17. A cubic block of concrete 4 ft. on a side is submerged in a tank of water which measures 6 ft. by 12 ft., and has a water depth of 10 ft. How much does the water level rise?

18. A log with a diameter of 14 inches and a length of 10 ft. weighs 40 lb./cu.ft. If it is inserted vertically into a body of water, what vertical force is required to hold it below the water surface?

19. A liquid with a Specific Gravity of 1.14 is pumped at a rate of 30 gpm. How many pounds per day are being delivered by the pump?

20. You take a core sample of your sand filter to check for percent mudball accumulation. The sampler is 6 inches in diameter and 12 inches long. After sifting out the sand, you immerse the mudballs into a 2 inch diameter graduate holding 1 liter of water. The water level in the graduate rises 4 inches. What is the percent mudballs in your filter?

21. A long narrow cylinder weighing 25 pounds was filled with water and placed on a scale. A 2 inch diameter probe hanging from the ceiling was submerged 3 feet in this water. If the scale registered 300 pounds, how many gallons of water were in the cylinder?

22. A cubical float, 4 ft. on a side, weighs 400 lb. and is anchored by means of a concrete block which weighs 1500 lb. in air. Nine inches of the float are submerged when the chain connected to the concrete is taut. What rise in the water level will lift the concrete off the bottom? (concrete density = 150 lb./cu.ft.)

SOLUTIONS

1. **Answer**   98.3 gallons

    Water weighs 8.34 lb./gal. Therefore:

    $$\frac{820\ lb.}{8.34\ lb./gal.} = 98.3\ gal.$$

---

2. **Answer**   8340 lb.

    Wt. Water = 8.34 lb./gal. x 1000 gal. delivered for a dollar = 8340 lb.

---

3. **Answer**   A. .7
                 B. 5.84 lb./gal.

    A. Density gasoline $= \dfrac{131\ lb.}{3\ cu.ft.} = 43.7$ lb./cu.ft.

    $$SG = \frac{density\ gasoline}{density\ water} = \frac{43.7\ lb./cu.ft.}{62.4\ lb./cu.ft.} = .7$$

    B. Multiply Specific Gravity by weight of a gallon of water.

        8.34 lb./gal x .7 = 5.84 lb./gal.

---

4. **Answer**   36.7 lb.

Solve for specific gravity; get lb./gal; multiply by 5.

$$SG = \frac{\text{wt. oil}}{\text{wt. water}} = \frac{55 \text{ lb./cu.ft.}}{62.4 \text{ lb./cu.ft.}} = .88$$

$$SG = \frac{\text{wt. oil}}{\text{wt. water}}$$

$$.88 = \frac{\text{wt. oil}}{8.34 \text{ lb./gal.}}$$

7.34 lb./gal. = wt. oil

7.34 lb./gal. x 5 gal. = 36.7 lb.

---

5. **Answer**   A. 41,700 lb. water
                    B. 76,311 lb. sulfuric acid

    A. Multiply by density of water.

        5000 gal. x 8.34 lb./gal = 41,700 lb. water

    B. Solve for density of one gallon sulfuric acid; multiply by 5000 gal.

$$SG = \frac{\text{density sulfuric acid}}{\text{density water}}$$

$$1.83 = \frac{\text{density sulfuric acid}}{8.34 \text{ lb./gal.}}$$

15.26 lb./gal. = density sulfuric acid

15.26 lb./gal x 5000 gal. = 76311 lb. sulfuric acid.

---

6. **Answer**   2.4

$$SG = \frac{\text{density concrete}}{\text{density water}} = \frac{150 \text{ lb./cu.ft.}}{62.4 \text{ lb./cu.ft.}} = 2.4$$

---

7. **Answer**   A. 848.64 lb./cu.ft.
                    B. 22914 lb./cu.yd.

    A. Multiply SG mercury by density water.

        13.6 x 62.4 lb./cu.ft. = 848.64 lb./cu.ft.

   B. Multiply density mercury by number cu.ft. per cu.yd.

       848.64 lb./cu.ft. x 27 cu.ft./cu.yd. = 22914 lb./cu.yd.

---

8. **Answer**   2700 lb.

Multiply by number of cu.ft. per cu.yd.

       100 lb./cu.ft. x 27 cu.ft./cu.yd. = 2700 lb.

---

9. **Answer**   A. .1875 lb./cu.ft.
                       B. 180 lb.

   A. Multiply by density of air (.075 lb./cu.ft.).

       .075 lb./cu.ft. x 2.5 = .1875 lb./cu.ft.

   B. Multiply density of chlorine by cu.ft. volume of room.

       Vol. = 10 ft. x 12 ft. x 8 ft.

       Vol. = 960 cu.ft.

       960 cu.ft. x .1875 lb./cu.ft. = 180 lb. chlorine in room

---

10. **Answer**   A. 47936 lb./day
                       B. 1.33

   A. Solve for lb. alum/gal.; multiply by gal./min. = lb./min.; multiply by 1440 min./day = lb./day.

$$\frac{83 \text{ lb./cu.ft.}}{7.48 \text{ gal./cu.ft.}} = 11.1 \text{ lb./gal.}$$

       11.1 lb./gal x 3 gal/min. pumped = 33.3 lb./min. pumped

       33.3 lb./min. x 1440 min./day = 47936 lb./day

   B. $SG = \dfrac{\text{density alum}}{\text{density water}} = \dfrac{83 \text{ lb./cu.ft.}}{62.4 \text{ lb./cu.ft.}} = 1.33$

11. **Answer**   A. 116532 lb.
                  B. 3.7 hrs.

     A.   Solve for bed volume (cu.ft.); solve for lb./cu.ft. sludge; multiply.

           Vol. bed = 30 x 60 x .83 = 1494 cu.ft.

$$SG = \frac{\text{wt. sludge}}{\text{wt. water}}$$

$$1.25 = \frac{\text{wt. sludge}}{62.4 \text{ lb./cu.ft.}}$$

           78 lb./cu.ft. = wt. sludge

           78 lb./cu.ft. x 1494 cu.ft. = 116532 lb. sludge applied to bed

     B.   Convert vol. bed to gallons; divide by 50 gpm.

           1494 cu.ft. x 7.48 gal./cu.ft. = 11175 gal. needed to fill bed

$$\frac{11175}{50 \text{ gpm}} = 223.5 \text{ min.} = 3.7 \text{ hrs.}$$

---

12. **Answer**   It will sink all the way to the bottom.

---

13. **Answer**   A. .64 cu.ft.
                  B. 2.25

     A.   The stone displaces 40 lb. of water: solve for cu.ft. of water displaced (equivalent to volume of stone).

$$\frac{40 \text{ lb. water displaced}}{62.4 \text{ lb./cu.ft.}} = .64 \text{ cu.ft. water displaced}$$

     B.   Stone volume = .64 cu.ft.; this weighs 90 lb.

$$\frac{90 \text{ lb.}}{.64 \text{ cu.ft.}} = 140.6 \text{ lb./cu.ft. density of stone}$$

$$SG = \frac{\text{density stone}}{\text{density water}} = \frac{140 \text{ lb./cu.ft.}}{62.4 \text{ lb./cu.ft.}} = 2.25$$

14. **Answer**   2.19 cu.ft.

This person will displace his total volume since his Specific Gravity is greater than that of water.

Solve for his density; then solve for the volume of person.

$$SG = \frac{\text{density person}}{\text{density water}}$$

$$1.1 = \frac{\text{density person}}{62.4 \text{ lb./cu.ft.}}$$

$$68.64 \text{ lb./cu.ft.} = \text{density person}$$

$$\frac{150 \text{ lb. person}}{68.64 \text{ lb./cu.ft. density person}} = 2.19 \text{ cu.ft. volume of person}$$

This person will displace 2.19 cu.ft. of water.

---

15. **Answer**   376 gallons

Solve for volume of coiled pipe, and volume of tank. Subtract pipe volume from tank volume. Change to gallons.

Pipe
Vol. = .785 d² x L
Vol. = .785 x .33² x 160
Vol. = 13.7 cu.ft.

Tank
Vol. = l x w x h
Vol. = 4 x 4 x 4
Vol. = 64 cu.ft.

64 cu.ft. - 13.7 cu.ft. = 50.3 cu.ft.

50.3 cu.ft. x 7.48 gal./cu.ft. = 376 gallons needed to fill tank.

---

16. **Answer**   49920 lb.

To submerge this box, it must become the same weight as an equivalent volume of water. Solve for that weight.

Vol. = l x w x h
Vol. = 10 x 10 x 8
Vol. = 800 cu.ft.

800 cu.ft. x 62.4 16/cu.ft. = 49,920 lb.

17. **Answer**   10.7 inches

Solve for volume of concrete block. This is the volume of displaced water. Apply to tank area to get depth of displacement.

Concrete Vol. = l x w x h
Concrete Vol. = 4 x 4 x 4
Concrete Vol. = 64 cu.ft.

Displacement Vol. = l x w x h
64 = 6 x 12 x h
.889 ft. = h
10.7 in. = h

---

18. **Answer**   240 lb.

If this log had the same weight as water, it would rest just barely submerged. Find the difference between its weight and that of the same volume of water. That is the weight needed to keep it down.

62.4 lb./cu.ft. (water)
- 40.0 lb./cu.ft. (wood)
22.4 lb./cu.ft. difference

Vol. wood = .785 x 1.167^2 x 10 ft. = 10.69 cu.ft.

22.4 lb./cu.ft. x 10.7 cu.ft. = 240 lb. needed to hold it below surface

---

19. **Answer**   410,400 lb./day

Solve for lb. pumped/min.; change to lb./day.

8.34 lb./gal. water x 1.14 SG liquid = 9.5 lb./gal. liquid

30 gal/min x 9.5 lb./gal = 285 lb./min.

285 lb./min. x 1440 min./day = 410,400 lb./day

---

20. **Answer**   3.7%

Solve for volume of sample; solve for volume of mudballs; solve for % mudballs.

Volume sampler = .785 x $.5^2$ x 1

Vol. = .19625 cu.ft.

Mudball volume = volume rise of water level in graduate

Volume rise = .785 x $.167^2$ x .33

Vol. = .00726 cu.ft.

$$\% \text{ mudballs} = \frac{.00726 \text{ cu.ft.}}{.19625 \text{ cu.ft.}} = .037 = 3.7\%$$

Percent mudballs in sampler = Percent mudballs in filter

---

21. **Answer**   32.5 gal.

The probe, even though suspended, registers weight on the scale - the weight of an equivalent volume of water.

Subtract weight of cylinder; solve for volume of immersed portion of probe; obtain weight of equivalent volume of water; subtract from total weight; change to gallons.

```
  300 lb. weight registered on scale
-  25 lb. weight of cylinder
  275 lb. weight of water plus immersed probe
```

Volume probe = .785 x .167^2 x 3

Vol. = .067 cu.ft.

Wt. of equiv. vol. of water = 62.4 lb./cu.ft. x .067 cu.ft. = 4.12 lb.

```
  275 lb. weight of water plus probe
- 4.12 lb. probe
  270.9 lb. weight of water
```

$$\frac{270.9 \text{ lb.}}{8.34 \text{ lb./gal.}} = 32.5 \text{ gal. water}$$

---

22. **Answer**   2.7 inches

Draw Diagram.   Weight of concrete block = underwater weight minus upward lift from attached float. Solve for underwater weight of concrete. Solve for upward lift from float. Subtract both of these from weight of concrete in air to obtain additional lift needed to get block to rise off bottom. Obtain upward lift for each inch of submersion depth of float; divide into total needed lift to get inches rise needed to lift block off bottom.

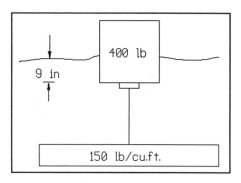

$$\frac{1500 \text{ lb. weight of concrete in air}}{150 \text{ lb./cu.ft. density concrete}} = 10 \text{ cu.ft. of concrete block}$$

62.4 lb./cu.ft. weight of water x 10 cu.ft. of water = 624 lb. buoyant force.

1500 lb. wt. concrete in air
-624 lb. buoyant force
876 lb. wt. of concrete underwater

Lift from float - all of float would be above water level if it were not held down. It is pulling the concrete block up with a force which is equivalent to the difference between the wt. of the submerged part and the weight of the same volume of water.

Volume float = 4 x 4 x 4 = 64 cu.ft.

$$\frac{400 \text{ lb. wt. of float}}{64 \text{ cu.ft.}} = 6.25 \text{ lb./cu.ft.}$$

Volume submerged float section = 4 x 4 x .75 = 12 cu.ft.

62.4 lb./cu.ft. density water x 12 cu.ft. submerged float = 749 lb.

6.25 lb./cu.ft. density float x 12 cu.ft. submerged float = 75 lb.

749 lb.
-75 lb.
674 lb. lift on concrete block from float

Therefore:

876 lb. wt. of concrete underwater
-674 lb. lift from float
202 lb. actual weight of concrete block attached to float

$$\frac{674 \text{ lb. total float lift}}{9 \text{ inches submerged float}} = 75 \text{ lb. lift for each inch of submersion}$$

$$\frac{202 \text{ lb. lift needed}}{75 \text{ lb. lift per inch water rise}} = 2.7 \text{ in. water rise to lift concrete off bottom}$$

CHAPTER 3

FLOW

Flow - the quantity of water passing a point in a given unit of time. Think of flow as a volume - which is moving. It can be recorded as gallons/day (gpd), gallons/minute (gpm), or cubic feet/second (cfs). Flows treated by a municipal utility are large, and often referred to in Million Gallons per Day (MGD).

Flow may be enclosed in pipes, or it may be open channel flow. The first 9 chapters of this text will be considering predominantly closed pipe flow. Pipes are flowing full, and under pressure. Open channel flow deals with conduits which are partially full, and the water surface is in contact with atmospheric pressure (streams, aqueducts, sewer pipes, grit chambers, clarifiers). The water flows because of the slope of the conduit. Open channel flow will be taken up later in the text.

Flow may be laminar or turbulent. Laminar flow only occurs at extremely low velocities. The water moves in straight parallel lines, called laminae, or streamlines, which slide upon each other as they travel, rather than mixing up. Turbulent flow, which is normal pipe flow, occurs because of friction encountered on the inside of the pipe. This throws the outside layers of water into the inner layers; the result is that all the layers mix and are moving in different directions, and at different velocities. Added up, however, the direction of flow is forward.

Flow may be steady or unsteady. We will be considering steady state flow only. At any one point, the flow, and velocity, does not change. Most hydraulic calculations are done under the assumption of steady state flow. If we can assume a given flow at a given point, calculations can be based on that fact. If we had to consider that the flow was constantly changing, calculations would become extremely complex. Note, however, that in actual systems, flow is frequently in an unsteady condition. As a water tank empties into a pipe, the flow through the pipe decreases over time, because the depth of water in the tank which pushes the water forward, is decreasing. As a pump fills a tank from a bottom entrance, that pump is delivering less water as the tank gets fuller and fuller. Flow through a wastewater treatment plant is constantly fluctuating, but process treatment units are designed for an average, maximum or minimum flow, depending on the need. We realize that unsteady flow is a common condition, but we do not bring this into calculations.

EQUATION OF CONTINUITY

Under the assumption of steady state flow, the flow that enters the pipe is the same flow that exits the pipe. It is continuous. Water is incompressible; it cannot accumulate inside. The flow at any given point is the same flow at any other given point on the pipeline. This is true in any water system, as long as no additional flows are added, and no exits which split or divert the flow away in other directions are present.

However, the velocity of the water may change. At a given flow, the velocity is dependent upon the cross sectional area of the conduit. Velocity is the speed at which the flow is traveling. A given quantity of water will travel faster through smaller spaces, and slower through larger spaces. Consider this phenomenon in a closed pipe. A 50 gal./min. flow travels at a very slow velocity through a 36 inch diameter pipe, but if the pipe narrows to 4 inches diameter, that same 50 gal./min. will be moving very fast through the narrow section, in order for the flow to remain the same.

The principle applies equally to open channel flow. Water entering a wastewater treatment plant travels at scouring velocity coming down the pipe (2-3 ft./sec.). Entering the grit chamber, which is larger in cross sectional area than the sewer pipe, the velocity slows down to about 1 ft./sec. to allow grit to settle. The next process unit, the primary clarifier, again takes advantage of this principle, and is large enough to slow the velocity to about .05 ft./sec. so that organics will settle. Much of wastewater treatment is designed upon this principle, and detention times are developed from it.

Hence was developed the Equation of Continuity:

$$Q = A V$$

Where:

> Q = Flow
> A = Area (cross sectional area of conduit)
> V = Velocity

This is the most basic hydraulic equation. It always holds true, and all other hydraulic formulas determine components of this one.

Calculations are very straightforward, solving for the unknown factor. Be aware that if pipes of different diameters are considered, substitution within the formula can be performed.

$\rightarrow$ flow in - equals - flow out $\rightarrow$

$$Q_1 \quad = \quad Q_2$$

Therefore:

$$A_1 \, V_1 \quad = \quad A_2 \, V_2$$

A calculation for velocity, or for pipe diameter, can often be done more quickly using this version of the formula.

Note that the unit labels used must be in the same dimensions across the formula. It is easiest to label flow as cu.ft./sec. (cfs), area as sq.ft., and velocity as ft./sec.

If values are presented in other dimensions, first change them to these before inserting into the formula.

PROBLEMS

1. A flow of 1 MGD occurs in an 8 inch diameter pipeline. What is the water velocity?

2. What is the velocity of water in a 36 inch diameter pipe which is carrying 50 gpm?

3. A city requires a flow of 40 MGD. What diameter pipe is required to carry this flow, if the water velocity is to be 4 ft./sec.?

4. The cross sectional measurements of a flume are 10 inches by 10 inches. The water is 8 inches deep, and moving at a velocity of 2 ft./sec. How many gallons of water will the flume deliver in four hours?

5. What capacity pump would be necessary in a sewage lift station to maintain a velocity of 3 ft./sec. in the 4 inch force main discharge pipe?

6. The velocity of flow in a 12 inch diameter pipe is 100 ft./min. The flow discharges through a nozzle 3 inches in diameter. What is the velocity of flow in the nozzle?

7. A rectangular open channel is 5 ft. wide and carries a flow of 12 MGD. A small child places his sailboat in the channel, and it travels a mile in 30 minutes. How deep is the channel?

8. How many gallons of water can be stored in a pipeline 5 ft. in diameter and 4 miles long?

9. A channel 4 ft. wide has water flowing to a depth of 2 ft. What is the gpm flow in the channel if the water travels at a speed of 3 ft./sec.?

10. A rectangular channel 24 inches wide and 18 inches deep is flowing half full. What is the discharge (cfs) when the velocity is 12 ft./min.?

11. An 8 inch diameter pipeline carries water to an industry at a velocity of 10 ft./sec.
    A. What is the flow in gpd?
    B. If water is purchased at $1.00/1000 gal., what is the quarterly bill for water usage?

12. A 24 inch diameter pipe with a water velocity of 60 ft./min. discharges into a rectangular clarifier 10 ft. deep, 10 ft. wide and 100 ft. long. What is the water velocity in the clarifier?

13. Flow = 10 MGD
    Find velocities A, B, C

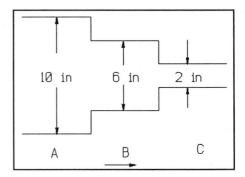

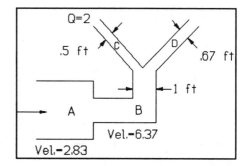

14. Presented is part of a pipe system.
    (Velocities are ft./sec.; Q = cfs)
    A. What is the diameter of pipe A?
    B. What is the velocity in pipe C?
    C. What is the flow in pipe D?

15. A 100 gpm pump:
    A. Will pump how many cfs?
    B. Will pump how many cu. meters/day?

16. A rectangular channel constricts from a width of 3 ft. to a width of 2.5 ft. in a short transition section. If the flow is 6.5 cfs and the depth upstream of the constriction is .4 ft., what is the depth downstream of the constriction.

17. A pipe 12 inches in diameter reduces to a diameter of 6 inches, then expands to a diameter of 10 inches. If the average velocity in the 6 inch diameter pipe is 15 ft./sec., what is the average velocity at the other sections?

18. A chlorine contact tank has a capacity of 6000 gal. and a length of 20 feet. Calculate the velocity in the tank when the flow is .432 MGD.

19. A 6 inch diameter pipe carries 2.87 cfs. The pipe branches into a 2 inch diameter pipe and a 4 inch diameter pipe. If the velocity in the 2 inch diameter pipe is 40 ft./sec., what is the velocity in the 4 inch diameter pipe?

20. Water flows through your trickling filter at a rate of 2 MGD. A portion of the water leaving the filter is being recirculated back to the filter, flowing through an 8 inch diameter pipe at a velocity of 2 ft./sec. What is the flow entering the headworks of this plant?

SOLUTIONS

1. **Answer**   4.4 ft./sec.

   Change MGD to cfs, inches to ft.; solve for velocity

   $$Q = A \ V$$

   $$1.55 = .785 \times .67^2 \ V$$

   $$4.4 = V$$

   $$4.4 \text{ ft./sec.} = V$$

2. **Answer**   .016 ft./sec.

Change gpm to cfs, inches to ft.; solve for velocity.

$$Q = A \quad V$$

$$.11 = .785 \times 3^2 \; V$$

$$.016 \; \text{ft./sec.} = V$$

---

3. **Answer**   4.5 ft. (54 inches)

Change MGD to cfs; solve for diameter.

$$Q = A \quad V$$

$$62 = .785 d^2 \times V$$

$$4.44 \; \text{ft.} = d$$

Choose the next larger pipe size (4.5 ft.).

---

4. **Answer**   118,333 gal./4 hrs.

Height of flume not needed. Change inches to ft.; solve for Q; change to gal./hr., then gal./4 hr.

$$Q = A \quad V$$

$$Q = .83 \times .67 \times 2$$

$$Q = 1.1 \; \text{cfs}$$

$$Q = .71 \; \text{MGD}$$

$$Q = 710,000 \; \text{gpd}$$

$$Q = 29583 \; \text{gal./hr.}$$

$$Q = 118,333 \; \text{gal./4 hrs.}$$

---

5. **Answer**   .257 cfs

Change inches to ft.

$$Q = A \quad V$$

$$Q = .785 \times .33^2 \times 3$$

$$Q = .257 \; \text{cfs}$$

6. **Answer**   26.7 ft./sec.

Change velocity to ft./sec.; assuming flow is the same in both pipes (steady state), can use $AV = AV$ and solve for velocity directly.

12 inch pipe	**3 inch nozzle**

$$A \quad V \quad\quad = A \quad V$$

$$.785 \times 1^2 \times 1.67 = .785 \times .25^2 \times V$$

$$26.7 \text{ ft./sec.} \quad = V$$

---

7. **Answer**   1.3 ft.

Change MGD to cfs; convert velocity to ft./sec.; solve for depth.

$$\frac{1 \text{ mile}}{30 \text{ min.}} = \frac{5280 \text{ ft.}}{1800 \text{ sec.}} = 2.93 \text{ ft./sec.}$$

$$Q = A \quad V$$

$$18.6 = 5 \times \text{depth} \times 2.93$$

$$1.3 \text{ ft.} = \text{depth}$$

---

8. **Answer**   3,100,000 gallons

Solve for volume of pipe.

$$\text{Volume} = .785 \times 5^2 \times 5280 \times 4$$

$$\text{Vol.} = 414480 \text{ cu.ft.}$$

$$\text{Vol.} = 3,100,000 \text{ gal.}$$

---

9. **Answer**   A. 10771.2 gpm

$$Q = A \quad V$$

$$Q = 4 \times 2 \times 3$$

$$Q = 24 \text{ cfs}$$

$$Q = 10771.2 \text{ gpm}$$

10. **Answer**   .3 cfs

Change inches to ft.; change ft./min. to ft./sec.

$$Q = A \ V$$

$$Q = 2 \times .75 \times .2$$

$$Q = .3 \text{ cfs}$$

---

11. **Answer**   A. 2,270,000 gpd
                  B. $204,300.00/quarter

A. Solve for flow; change to gpd.

$$Q = A \ V$$

$$Q = .785 \times .67^2 \times 10$$

$$Q = 3.52 \text{ cfs}$$

$$Q = 2,270,000 \text{ gpd}$$

B. Multiply gpd by days/quarter; convert to units of 1000 gal.; multiply by $.

2,270,000 gpd x 90 days/qtr. = 204,300,000 gal./qtr.

$$\frac{204,300,000 \text{ gal./qtr.}}{1000} = 204,300 \times \$1 = \$204,300/\text{qtr.}$$

---

12. **Answer**   .03 ft./sec.

Assume steady state flow through pipe and clarifier. Use AV=AV; change inches to ft. and velocity to ft./sec.

	Pipe	**Clarifier**

$$A \times V \ = A \times V$$

$$.785 \times 2^2 \times 1 = 10 \times 10 \times V$$

$$.03 \text{ ft./sec.} \quad = V$$

---

13. **Answer**   A. 28.7 ft./sec.
                  B. 78.9 ft./sec.
                  C. 707.8 ft./sec.

Changes inches to ft., MGD to cfs.

A. $$Q = A \ V$$

$$15.5 = .785 \times .83^2 \times V$$

$$28.7 \text{ ft./sec.} = V$$

B. $Q = A \ V$

                $15.5 = .785 \times .5^2 \times V$

                $78.9 \ ft./sec. = V$

C. $Q = A \ V$

                $15.5 = .785 \times .167^2 \times V$

                $707.8 \ ft./sec. = V$

---

14. **Answer**   A. 1.5 ft.
                       B. 10.2 ft./sec.
                       C. 3 cfs

A. Steady state flow through A & B; solve for diameter.

         $A \ V = A \ V$

         $.785 \times d^2 \times 2.83 = .785 \times 1^2 \times 6.37$

         $d = 1.5 \ ft.$

B. Dimensions are given; solve for velocity.

         $Q = A \ V$

         $2 = .785 \times .5^2 \ V$

         $10.2 \ ft./sec. = V$

C. Solve for flow in pipe B; subtract flow in C = flow in D.

         $Q = A \ V$

         $Q = .785 \times 1^2 \times 6.37$

         $Q = 5 \ cfs$

         5 cfs (B) - 2 cfs (C) = 3 cfs (D)

---

15. **Answer**   A. .22 cfs
                       B. 545 cu.meters/day

A. Convert 100 gpm to cfs.

B. Convert 100 gpm to gpd, then to L/day, then to cu.meters/day.

---

16. **Answer**   .4 ft.

Depth is a component of cross sectional area. Flow is steady, velocity at the constriction increases because of width decrease. Assuming that downstream of constriction velocity and width return to normal, depth should remain constant.

17. **Answer**   A. 12 inch section velocity = 3.75 ft./sec.
B. 10 inch section velocity = 5.45 ft./sec.

Draw Diagram.

A. Change inches to ft.; using AV=AV, solve for vel. 12 inch section.

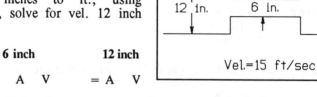

6 inch	12 inch
A V	= A V

$.785 \times .5^2 \times 15 = .785 \times 1^2 \times V$

3.75 ft./sec.   = V

B. Same method, solve for velocity in 10 inch section.

6 inch	10 inch
A V	= A V

$.785 \times .5^2 \times 15 = .785 \times .83^2 \times V$

5.45 ft./sec.   = V

18. **Answer**   .0167 ft./sec

Change MGD to cfs and gal. to cu. ft.; using volume, calculate area of tank.

Volume tank = Area x Length

802.1 = A x 20

40.1 sq.ft. = A

Q = A V

.67 = 40.1 x V

.0167 ft./sec. = V

19. **Answer**   23.3 ft./sec.

Draw Diagram. Change inches to ft.;
solve for flow in 2 inch diameter pipe,
and subtract from flow in 6 inch
diameter pipe = flow in 4 inch
diameter pipe. Solve for velocity in 4
inch pipe.

Vel.=40 ft/sec

2 in.

2.87 cfs   6 in.   4 in.

$Q = A \ V$

$Q = .785 \times .167^2 \times 40$

$Q = .88 \ cfs$

2.87 cfs (6 inch) - .88 cfs (2 inch) = 1.99 cfs (4 inch)

$Q = A \ V$

$1.99 = .785 \times .33^2 \times V$

$23.3 \ ft./sec. = V$

20. **Answer**   2.4 cfs

Draw Diagram. Change MGD to cfs,
inches to ft.; solve for recirc. flow,
subtract from TF flow.

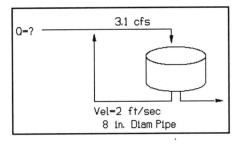

Q=?   3.1 cfs

Vel=2 ft/sec
8 in. Diam Pipe

$Q = A \ V$

$Q = .785 \times .67^2 \times 2$

$Q = .7 \ cfs$ - being
        recirculated

Plant flow + Recirc. flow =
Trickling filter flow

$X + .7 \ cfs = 3.1 \ cfs$

$X = 2.4 \ cfs$ - incoming plant flow

CHAPTER 4

PRESSURE

As all fluids, water has a specific weight: 62.4 pounds for every cubic foot, or 8.34 pounds for every gallon. This weight, resting on a surface, exerts a force on that surface. For example, one cubic foot of water, resting on its bottom surface, exerts a force of 62.4 pounds on that square foot (62.4 lb./sq.ft.). Force on a unit area is called <u>Pressure</u>.

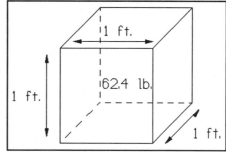

Note here that the unit label for specific weight (density), is a weight per volume designation (lb./cu.ft.).

The unit label for Pressure is a weight per area designation (lb./sq.ft.). It applies to the surface area that the weight is resting on.

If our weight were composed of two cubic foot blocks of water, one on top of each other, then the specific weight would be 62.4 lb./cu.ft., the height would be 2 feet, and the pressure on the bottom surface would be 124.8 lb./sq.ft. Density remains constant, but the pressure on the unit area depends on the height, or depth of the water.

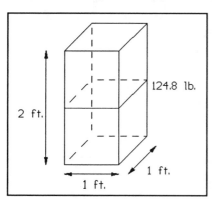

A formula develops:
Pressure = Weight x Height
124.8 lb/sq.ft. = 62.4 lb/sq.ft. x 2 ft.

If we had four blocks of water, arranged as in the diagram below, the pressure on every square foot of bottom surface would still be 124.8 lb./sq.ft. It would not matter how wide the water was. It is just the height, or depth, that is pushing down on that surface, and causing the pressure.

It is most convenient to specify pressure at a particular point in a water system. For example, what is the water pressure at a specified point along a pipeline? Or, what is the pressure at a particular point on the bottom of a swimming pool? Using the concept of pressure, we now need to deal with a very small unit area. Therefore, pounds per square inch (psi) has become the favored designation.

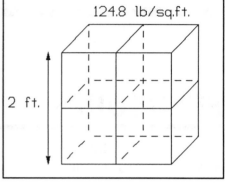

Let's take another look at our cubic foot block of water. Wishing to register pressure as lb./sq.in., we must split that bottom surface into 144 square inches. Now, how much water is resting on each of those square inches? Picture a column of water rising from each of them to the top of the block. How much does each column weigh?

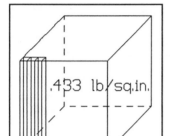

$$\frac{62.4 \text{ lb./cu.ft.}}{144 \text{ sq.in./sq.ft.}} = .433 \text{ lb. per foot of height}$$

A cubic foot block of water, one foot high, exerts a pressure of .433 lb. on every square inch of bottom surface area.

Our double block of water, two feet high, exerts a pressure of .866 lb. on every square inch of bottom surface area. In all cases, the pressure depends upon the depth of the water above it, and an equivalence of feet of depth to psi pressure is created.

However, there is a more comfortable way to state this equivalence. For practical purposes, the water utility superintendent would like to know how many feet of water depth he has to provide in the city water storage tank in order to have adequate pressure in the pipes. There is needed an equivalence of feet of water depth to psi pressure that is readily interchangeable, and which will give us discrete units of psi pressure.

Creating a ratio, if a 1 ft. column of water is equivalent to a pressure of .433 psi, then how many feet of water will give us 1 psi?

$$\frac{.433 \text{ psi}}{1 \text{ ft.}} = \frac{1 \text{ psi}}{x \text{ ft.}}$$

$$x = 2.31 \text{ ft.}$$

Therefore, we see that 1 psi is equivalent to a 2.31 ft. depth of water. Water pressure can be registered as psi, or as feet of water. The conversion 1 psi = 2.31 ft. is most commonly used. However, .433 psi = 1 ft. will give the same result in calculation.

PASCAL'S LAW

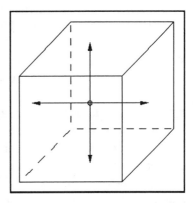

In the 17th century, Blaise Pascal formulated a fundamental law of hydraulics. The pressure at any one point in a static liquid is exerted with equal intensity in all directions. For instance, the pressure at any point on the horizontal bottom of a tank of water is the same, and the pressure exerted upward, downward, and to the sides from any one point is the same also. Dewatered basement floors in flooded areas buckle from this upward pressure, and structures built underground must be able to withstand upward pressures imposed by groundwater, or they may float right up out of the ground.

CONTAINER SHAPE

Now take both of these facts:

Pressure in a static system is due to the depth of water above the point of measurement.

Pressure is exerted with equal intensity in all directions.

The source of distribution system pressure in most community water systems is the water storage tank. These tanks may take a number of shapes, but these same basic hydraulic principles apply. In each of the diagrams below, as long as the depth of the water is the same from the water surface to the point of measurement, the water pressure at that point is the same. The shape of the tank does not matter, nor does the number of gallons each holds.

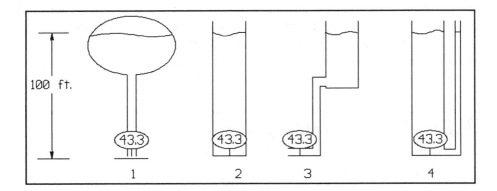

#1 is a typical elevated tank. Let us say that it holds 400,000 gallons of water when the water level is 100 feet above ground surface. A pressure gage at ground level will register 43.3 psi, because of the 100 ft. depth of water above it. The pressure registers downward from the water surface, straight through the riser pipe to the gage.

#2 is a standpipe, also containing 400,000 gal. when the water surface is 100 ft. above ground. It will also register 43.3 psi at the bottom. The difference between these two types is in the use of the water. The water pressure which supplies the community from #1 will remain high until the tank is empty. #2 provides the same pressure when full, but as water is used and the level drops, the pressure will drop significantly with it.

#3 is a ground level storage tank, set into a hill, with a pressure gage on the pipeline 100 ft. below the water surface. The pressure gage will register 43.3 psi. The community is down at this level, and the water storage tank was installed at a higher elevation to provide the pressure. It doesn't matter that the tank is buried in the ground instead of being up in the air. The same depth of water is still provided by the direct physical connection of water from the tank through the pipeline. The pressure registers down, and to the side, all the way down the pipe to the gage location.

#4 is a copy of #2, equipped with a ½ inch diameter sight glass, to read the water depth. Connected to the bottom of the tank, the water will rise in the glass to the water level in the tank - 100 ft. The gage at the bottom of the tank registers 43.3 psi. If a valve between the sight glass and the tank is then closed, a gage at the bottom of the sight glass will still register 43.3 psi. Though there is only a little water in that glass, it is 100 ft. deep above the gage.

PUMPS

Adequate elevation and depth of water in a tank will provide enough pressure to move the water throughout the distribution system.

PUMPS ALSO PROVIDE PRESSURE.

When elevation is not available, a pump is used to provide the needed pressure. The pump is actually providing lift (elevation), or enough pressure to raise the water to a given height. If the direction of flow is horizontal from the pump discharge, then the pressure remains in the discharge pipes, and keeps the water moving. Pumped pressure can also be registered in feet of water, or in psi.

PRESSURE MEASUREMENT
Piezometer

The simplest type of pressure measurement device is the piezometer. This is simply a narrow, open-ended vertical tube, attached at the point of pressure measurement. Water will rise in the tube to a height equivalent to the pressure. If a piezometer is attached to the bottom of a tank, and is made of glass, it is referred to as a sight glass, and provides an easy means of reading the water level. The water depth can easily be converted to psi to obtain a pressure reading at the bottom. This is practical for small

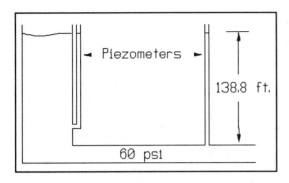

tanks of water at ground level, but obviously impractical on an elevated storage tank. Piezometers can be installed on pipelines. The water will rise in the tube to a level equivalent to the pressure in the pipe. As long as the piezometer is tall enough, the water will not pour out the top. It will rise till the depth of water in the tube provides a downward pressure equivalent to the upward pressure in the pipe. Again, this is most often impractical. A pipeline under a pressure of 60 psi, would need a piezometer 138.6 ft. high.

Pressure Gage

The pressure gage is a compact, practical device used for pressure measurement. It is probably the most important instrument in the water system. Called a <u>Bourdon Tube</u>, it should be calibrated in feet of water, though it may read feet, psi, or inches of mercury. The dial size is usually 4½ inches in diameter, for readability, and the range selected should go beyond the maximum operating pressure of the system. The needle can be zeroed with a screwdriver. The gage assembly is a liquid filled system. The snubber is a restrictive device that stops movement of the liquid fill, deadening pulses of power which cause a blurred pointer motion. Glycerine is often preferred for fill, and prevents solids bearing water from damaging the gage. The diaphragm seal isolates the water from the fill, but allows the pressure to be transmitted through. The bleed screw on the bottom of the diaphragm seal is for bleeding air if the pointer does not return to zero when the gage is disconnected. A ball valve should be at the bottom for shutoff. The nipple is the connection of the gage to the pipe, and should be welded on,

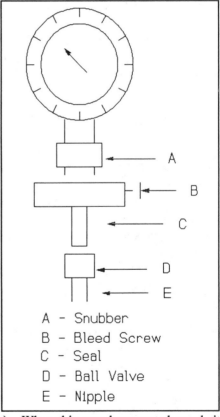

A – Snubber
B – Bleed Screw
C – Seal
D – Ball Valve
E – Nipple

or attached with a service saddle (not tapped in). When this area becomes clogged, it is easy to remove the gage above the valve, and rod out the nipple and valve.

GAGE POSITION

A pressure gage is usually attached directly onto the pipe, and reads pressure directly from it. Occasionally, for ease in reading, the gage is placed at some distance from the pipe, either above or below it. If an automatic correction for the difference in elevation is not built into the apparatus, it is easy to calculate. A gage placed below the pipeline will register a higher pressure than the pressure in the pipe. The water extends down to the diaphragm seal on the gage. If the extension is 5 ft. long, then the gage will register the pipe pressure, plus the extra five feet of elevation. If the gage is above the pipeline, it will register the pipe pressure minus the distance between the gage and the pipe. In this case the pipe pressure is more than the gage pressure.

Be careful to remember to change pressure units from psi to feet when calculating in these distances.

Keep in mind that even if the water in the pipe is flowing, the water exiting to the pressure gage is static. It dead ends at the gage; this water is not moving, and except for the difference in elevation from the pipe to gage, the actual pipe pressure at the point of connection is being transmitted to the gage. Bends and turns in the gage extension line will not affect the gage reading (Pascal's law-pressure registers in all directions). Only the elevation difference between gage and pipeline is significant.

ABSOLUTE PRESSURE

The pressure read on a pressure gage is referred to as Gage Pressure. It is the difference between a given pressure and that of the atmosphere. It is most often a positive number, but can also be negative, such as in siphon lines, or on the suction side of a pump which draws water up from below. Some gages are calibrated to read negative pressures, and register it with a minus sign. These gages are reading the amount of pressure below atmospheric pressure. They are actually reading the amount of vacuum in the line.

Atmospheric Pressure

The atmosphere is composed of gases. They have weight, as liquids do, and exert a pressure upon the earth. We don't feel this pressure, because we are so used to walking around in it, but the atmosphere, about 200 miles deep, exerts a pressure of approximately 14.7 psi at sea level, the standard reference point for all pressure measurements. It exerts that same pressure on an open tank of water. In fact, it is this pressure that keeps water in a liquid state. If a tank of water were sealed, and all the air pressure vacuumed out of it, the water would become vapor. It would no longer have the force upon it which holds it together as a liquid. We see some evidence of this on days when a zone of low atmospheric pressure enters our area; water evaporates more easily, creating cloudy, humid days. Weather stations record atmospheric pressure on a barometer, a piezometer filled with mercury (atmospheric pressure = 30 inches mercury). If we could vacuum all the air out of a container, the pressure reading on an attached gage would be -14.7 psi, or -34 feet of water: Absolute Vacuum. This is equivalent to an Absolute Pressure reading of 0.

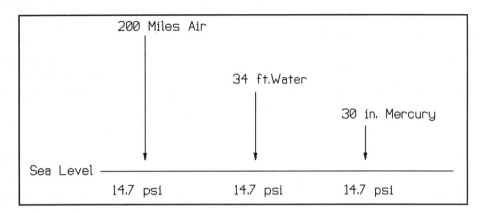

Absolute Pressure

Pressure readings can be reported either as Absolute Pressure or as Gage Pressure. The difference is that in Gage Pressure readings, atmospheric pressure is not counted. With Absolute Pressure readings, it is.

$$\text{Absolute Pressure (psia)} = \text{Gage Pressure (psig)} + \text{Atmospheric Pressure (14.7 psi)}$$

Absolute Pressure is always positive. It is theoretically impossible to have an Absolute Pressure reading below zero.

Gage Pressure is the pressure above or below atmospheric, and is recorded as psig. It can be negative - down to minus 14.7. The Absolute System of measuring pressures is less commonly used in hydraulic calculations, and this text will deal only with Gage Pressure unless otherwise indicated.

TOTAL PRESSURE - FORCE

Force is the pressure on an entire area. The force downward on the entire bottom of a tank filled with water, laterally on the side of a dam, or upward on the cover of an anaerobic digester. Force is registered in pounds, and is calculated by multiplying the pressure (lb./sq.ft.) times the area (sq.ft.). The formula:

Force = Pressure x Area

lb. = lb/sq.ft. x sq.ft.

can be used with other unit labels (square inches) as well, as long as the same dimensional units are carried across the entire formula.

Our original cubic foot of water exerts a force of 62.4 pounds on its bottom surface of one square foot. This is the entire weight, the total pressure on that surface. The unit pressure (psi) is only .433, yet the force is much larger. A small pressure, when applied to a large area, can account for a large force. For example, the gas pressure in an anaerobic digester should not exceed 11 inches of water column, to avoid blowing out the water seal. Eleven inches of water column is only .4 psi. But the upward force on a digester cover 40 ft. in diameter at that pressure is over 72,000 pounds. When calculating force on a vertical surface, such as the side of a tank, an average must be taken. The pressure at the water surface is zero; the pressure at the bottom is maximum; therefore, an average of the two must be used, and then multiplied by the sidewall area.

Force also becomes a significant factor in treatment processes when dealing with process tanks which are set below ground surface. If the water table is high in the area, it may be necessary to dewater the area around the tank before dewatering the tank for cleanout or repairs, or the surrounding force of water exerted upward on the bottom of the empty tank will pop it right out of the ground.

When a force is applied to a gas, the gas is able to absorb the pressure, and compresses to a smaller volume. When the pressure is released, more space is available, and the gas expands to fill the entire space. For practical purposes, liquids do not have this capacity. They are incompressible.

WATER IS AN INCOMPRESSIBLE FLUID.

When pressure is applied, it will not condense; volume remains the same, and pressure is transmitted to the surroundings. This yields dramatic results when moving water comes to a sudden halt.

EFFECTS OF PRESSURE - DYNAMIC SYSTEMS
Water Hammer

Called Hydraulic Shock, water hammer is the momentary increase in pressure which occurs in a moving water system when there is a sudden change of direction or velocity of the water.

When water flowing in a pipeline is suddenly stopped by a rapidly closing valve, pressure energy is transferred to the valve and pipe wall. Shock waves are set up within the system. Waves of pressure travel backwards until encountering the next solid obstacle, then forward, then back again. Neither the pipe nor the water will compress to absorb the shock. Water is 100 times less compressible than steel; this sudden increase in pressure will damage pipes, valves, and shake fittings loose. The velocity of a pressure wave is equal to the speed of sound; therefore it "bangs" as it travels back and forth, until dissipated by friction losses. We are familiar with the "bang" that resounds throughout an older house when a faucet is suddenly closed - an effect of water hammer.

Air chambers are installed in areas where water hammer is encountered frequently, and are typically seen behind sink and tub fixtures. Shaped like thin upside-down bottles, with a small orifice connection to the pipe, they are air filled. The air compresses to absorb the shock, protecting the fixture and piping. On pumps, which can be damaged by hammer caused when electrical power fails, the best form of prevention is to have automatically controlled valves which close slowly.

A less severe form of hammer is called Surge, a slow motion mass oscillation of water caused by pressure fluctuations in the system. This could be pictured as a slower "wave" of pressure building within the system. In long pipelines, surge can be relieved with a surge tank, an open tank of water directly connected to the pipeline. A surge tank is quite large, and the water level in it will reflect the pressure in the pipeline. When surge is encountered, it will act to relieve the pressure, and can store excess liquid. Surge tanks can serve for both positive and negative pressure fluctuations.

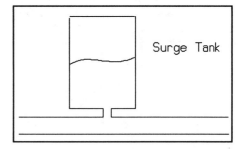

Both water hammer and surge are referred to as transient pressures. They both yield the same results if not controlled-damage to pipes, fittings, valves, causing leaks and shortening the life of the system.

Excess pressure in water lines can also be caused by entrained air, or by temperature changes of the water. Air trapped in the line will compress, and will exert extra pressure on the water. Temperature changes will actually cause the water to expand or contract, also affecting pressure. These conditions can be controlled by pressure relief valves, set to open with excess pressure in the line, and then close when pressure drops. Usually spring loaded, they are referred to as "surge suppressors". These are an integral part of a heating system, along with air chambers to hold excess water at high temperatures.

Thrust

Thrust is the force which water exerts on a pipeline as it turns a corner. The phenomenon is similar to water hammer, except that in a moving system, it is happening constantly. Caused by an imbalance of internal pressures, the intensity of thrust is dependent upon momentum (mass times velocity). The larger the pipeline, the faster the water is moving, the more the thrust. The force can be very great in large pipes, even when pressures are low. Thrust acts perpendicular to the outside corner of the pipe. It affects bends, tees, reducers, dead ends, and tends to push the coupling away from both sections of pipeline. Uncontrolled, thrust will separate the coupling, and cause the

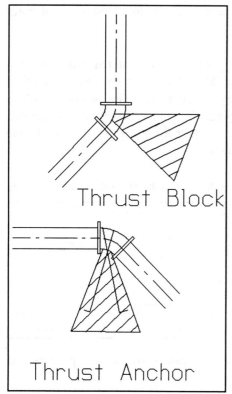

Thrust Block

Thrust Anchor

pipe to leak. Two devices are in use for thrust control in larger pipelines.

Thrust Blocks - a mass of concrete cast in place onto the pipe and around the outside corner of the turn. These are used for pipes with an elbow or tee which turns right or left or slants upward. The thrust force is transferred to the soil through the larger bearing area of the block.

Thrust Anchors - a restrained joint system employing a massive block of concrete cast in place below a downward bend and shackled to it with steel rods.

The size and shape of a thrust control device depends on pipe size, type of fitting, water pressure, water velocity, expected water hammer, and soil type. The block dimensions are calculated by determining the amount of thrust expected for the particular fitting based on pressure, velocity, size and type of fitting, then calculating the amount of thrust the particular soil will support per square foot of thrust block. In places where thrust blocks cannot be used (if the soil has poor bearing strength, or if there is no space for a massive block of concrete), restrained joint pipe is an alternative. The joint directly transfers the thrust load from the pipe to the surrounding soil.

PROBLEMS

1. A pressure gage at bottom of a standpipe reads 105 psi. What is the depth of water in the standpipe?

2. House elevation = 1520 ft.
   Gage A elevation = 2222 ft.
   Water surface elevation = 2390 ft.
   A. With valve A open and valve B closed, what does gage B read (psi)?
   B. With valve B closed and valve A closed, what is the reading at B?

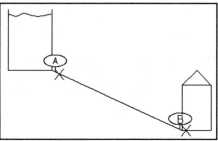

3. A 100 ft. diameter cylindrical tank contains 1.5 MG water.
   A. What is the water depth?
   B. What pressure does a gage at the bottom read (psi)?

4. The pressure in a pipe is 65 psi.
   A. What is the pressure in feet of water?
   B. What is the pressure in psf?

5. What height of a column of oil (SG=.84) would exert the same pressure as a column of water 46 ft. high?

6. What is the pressure of a column of mercury (SG=13.6) which is 10 ft. high, on one square inch of its bottom surface?

7. The pressure in a pipeline is 6234 psf. What is the head on the pipe?

8. A gage on the suction side of a pump shows a vacuum of 10 inches of mercury.
   A. What is this pressure in feet of water?
   B. What is the pressure in psi?

9. A pressure gage positioned 7.5 ft. above a pipeline reads 200 psi. What is the pressure in the pipe?

10. Existing 30 psi pressure in a pipeline is transmitted through 300 ft. of tubing to a gage which is 10 ft. below the pipe. What is the pressure reading on the gage?

11. What must the pressure in a water pipe 30 ft. below a closed faucet be to give a pressure of 35 psi at the faucet?

12. The pressure on a surface is 37 psig. If the surface area is 1.8 sq.ft. what is the force (lb) exerted on the surface?

13. A common vertical wall between two tanks of liquid measures 20 ft. high and 20 ft. wide. Water in one tank is 15 ft. deep. Oil (SG=.77) in the other tank is 6 ft. deep. What is the force acting on the wall?

14. A triangular trough carrying water is 200 ft. long. What is the total stress on the side of the trough if water in it is 3 ft. deep, and the width of the water surface is 8 ft.?

15. A concrete gravity dam with a vertical upstream face is 80 ft. high. What is the total force on the dam when the water level is 4 ft. below the top of the dam? Dam width at river bed is 30 ft. and sides slope at 45 degree angle outward vertically.

16. A water storage tank on top of a building supplies the cylindrical hot water heater in the basement, 50 ft. below. The water storage tank has 15 ft. of water in it. The hot water heater is 6 ft. tall and has a diameter of 4 ft. What is the average force on the heater walls?

17. A tank full of water measures 15 ft. long, 10 ft. wide and 12 ft. high.
    A. What will a pressure gage at the bottom read (psi)?
    B. What is the force of the water on the bottom?

18. A tank is 10 ft. by 15 ft. by 25 ft. water depth. What is the pressure at the bottom:
    A. in psig
    B. in psia
    C. in psfg
    D. in psfa

19. A closed rectangular steel box 40 ft. by 20 ft., and 20 ft. high, and weighing 80 tons, is to be sunk in water.
    A. How deep will it sink when launched?
    B. If the water is 18 ft. deep, what weight must be added to sink it to the bottom?

20. An iceberg has a specific weight of 57.2 lb./cu.ft. What portion of its total volume will extend above the surface if it is in fresh water?

21. A. A basement which measures 35 ft. by 60 ft. by 7 ft. high has its floor 6 ft. below grade. Assuming the total weight of the house and basement is 1,250,000 lb., what elevation of the groundwater would be critical for this house?
    B. The owner now paints the house with a "waterproof" paint which successfully resists a pressure equivalent to a 96 inch column of oil (SG=.80). Will the basement leak before it floats? Why?

SOLUTIONS

1. **Answer**   242.6 ft.

   Convert pressure to ft. of water.

   $$105 \text{ psi x } 2.31 \text{ ft./psi} = 242.6 \text{ ft.}$$

2. **Answer**   A. 376.6 psi
                 B. 304 psi

Draw Diagram.

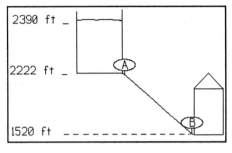

A. Gage B registers pressure from tank
    elevation, and from water depth.

>   2390 ft.
>   - 1520 ft.
>   870 ft.

870 ft. div. by 2.31 ft./psi = 376.6 psi

B. With both valves closed, gage B registers only the pressure in the pipe due to
    elevation difference. The water in the tank does not register.

>   2222 ft.
>   - 1520 ft.
>   702 ft. div. by 2.31 ft./psi = 304 psi

---

3. **Answer**   A. 25.6 ft.
                 B. 11.1 psi

A. Change MG to cu.ft.; using volume, solve for depth.

$$\frac{1,500,000 \text{ gal.}}{7.48} = 200535 \text{ cu. ft.}$$

Vol. = .785 x d^2 x depth

200535 = .785 x 100^2 x depth

25.6 ft. = depth

B. Convert to psi.

$$\frac{25.6 \text{ ft.}}{2.31} = 11.1 \text{ psi}$$

---

4. **Answer**   A. 150.2 ft.
                 B. 9361 psf

A. Convert pressure to ft. of water.

65 psi x 2.31 ft./psi = 150.2 ft.

B. Convert psi to psf.

65 psi x 144 sq.in./sq.ft. = 9361 psf

5. **Answer**   54.8 ft. of oil

The weight of oil is .84 of the weight of water.

$$\frac{46 \text{ ft. water}}{.84} = 54.8 \text{ ft. oil}$$

---

6. **Answer**   58.9 psi

Mercury is 13.6 times heavier than water.

.433 lb./ft./sq.in.(wt. water) x 10 ft. x 13.6 = 58.9 psi

---

7. **Answer**   100 ft. head

Head on pipe = ft. of pressure

Pressure = Weight x Height
6234 psf = 62.4 lb./cu.ft. x height
100 ft. = Height

---

8. **Answer**   A. -11.3 ft.
                 B. -4.9 psi

A. Change to inches of water; then to ft. water.

10" mercury x 13.6 (SG mercury) = 136" water = -11.3 ft. water

B. Change to psi.

$$\frac{-11.3 \text{ ft.}}{2.31 \text{ ft./psi}} = -4.9 \text{ psi}$$

This is a vacuum condition (less than atmospheric pressure). It occurs when pumps draw a suction lift.

---

9. **Answer**   203.2 psi

Draw Diagram. Convert pipe pressure to ft.; add for elevation; change back to psi.

200 psi x 2.31 = 462 ft.

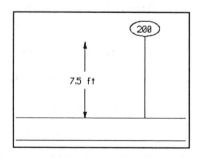

    462 ft.
+   7.5 ft.
   469.5 ft.

469.5 ft. div. by 2.31 = 203.2 psi

10. **Answer**   34.3 psi

Draw Diagram. Change pressure to ft.;
add elevation difference; change back to
psi. (The length of the line does not
matter.)

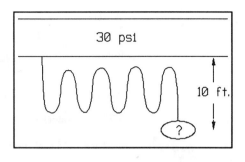

30 psi x 2.31 = 69.3 ft.

69.3 ft. + 10 ft. = 79.3 ft.
div. by 2.31 = 34.3 psi

---

11. **Answer**   48 psi

Draw Diagram. There must be more pressure
below. Change 35 psi to ft.; add elevation; change
back to psi.

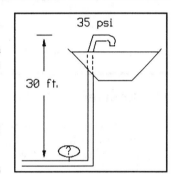

35 psi x 2.31 = 80.9 ft.

80.9 ft. + 30 ft. = 110.9 ft. needed
below; div. by 2.31 = 48 psi

It will take 48 psi, 30 ft. below the faucet to
produce 35 psi at faucet.

---

12. **Answer**   9590.4 lb. force

Covert psi to psf; solve for force.

37 psi x 144 sq. in./sq.ft. = 5328 psf

Force = Pressure x Area

lb. = 5328 psf x 1.8 sq. ft.

lb. = 9590.4

---

13. **Answer**   97200 lb. force

Draw Diagram. The two liquids apply
force in opposite directions. Find the
average pressure of each on the wall;
subtract one from the other; solve for
force.

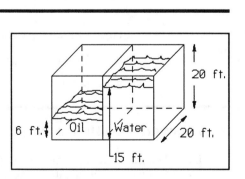

Oil:
Top pressure = 0

Bottom pressure = .433 lb./sq.in./ft. x .77 x 6 ft. = 2 psi

Average Pressure = 1 psi

Water:
Top pressure = 0

Bottom pressure = $\frac{15}{2.31}$ ft. = 6.5 psi

Average pressure = 3.25 psi

3.25 psi (water) - 1 psi (oil) = 2.25 psi pressure from the water side

2.25 psi x 144 sq.in./sq.ft. = 324 psf

Force = Pressure x Area

lb. = 324 psf x 20 ft. x 15 ft.

lb. = 97200

---

14. **Answer**   93500 lb. force

Draw Diagram. The force is along the entire 200 ft. of length, on the slanted side. Change ft. of water to psf.; solve for force.

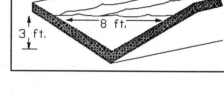

Pressure = Weight x Height

psf = 62.4 lb./cu.ft. x 3 ft.

psf = 187.2 at bottom

psf = 0 at top

psf = 93.6 average pressure

Side of triangle can be determined by Pythagorean theorem from geometry $(C^2 = A^2 + B^2)$.

Side = 5 ft. (This is a 3-4-5 right triangle.)

Force = Pressure x Area

lb. = 93.6 x 5 ft. x 200 ft.

lb. = 93500

---

15. **Answer**   19141056 lb. force

Draw Diagram. Set up dimensions; because of 45° slope, water depth is the same as width of water surface in the outer triangles (76 ft.); both triangles can be placed together to form a square 76 ft. x 76 ft.

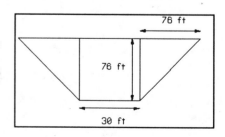

Solve for average pressure on the dam; solve for force.

Pressure at top $= 0$

Pressure at bottom $= \dfrac{76}{2.31}$ ft. $= 32.9$ psi

Average pressure $= 16.5$ psi x 144 sq.in./sq.ft. $= 2376$ psf

Solve for area.

Mid section: 76 ft. x 30 ft. $= 2280$ sq.ft.

Ends together 76 ft. x 76 ft. $= 5776$ sq.ft.

2280 sq.ft. $+$ 5776 sq.ft. $= 8056$ sq.ft.

Force $=$ Pressure x Area

lb. $= 2376$ psf x 8056 sq.ft.

lb. $= 19141056$ lb.

---

16. **Answer**   290,984 lb. force

Draw Diagram. Find average pressure on heater walls; solve for area of heater walls; solve for force.

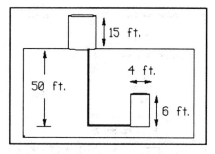

Pressure at top of heater $= 44$ ft. elevation to storage tank plus 15 ft. water depth in it $= 59$ ft.

$\dfrac{59}{2.31}$ ft. $= 25.5$ psi

Pressure at bottom of heater $= 15$ ft. $+ 50$ ft. $= \dfrac{65}{2.31}$ ft. $= 28.1$ psi

Average pressure $= 26.8$ psi

Average pressure $= 3859.2$ psf

Area of heater walls $=$ circumference x height

Area $=$ pi x diam. x height

Area $= 3.14$ x 4 x 6 ft.

Area $= 75.4$ sq.ft.

Force $=$ Pressure x Area

lb. $= 3859.2$ psf x 75.4 sq.ft.

lb. $= 290984$

17. **Answer**   A. 5.2 psi
                   B. 112320 lb. force

   A. Pressure is equivalent to depth.

   $$\frac{12 \text{ ft.}}{2.31} = 5.2 \text{ psi}$$

   B. Maximum pressure registers on tank bottom surface as force.

   5.2 psi x 144 sq.in./sq.ft. = 748.8 psf

   Force = Pressure x Area

   lb. = 748.8 psf x 15 ft. x 10 ft.

   lb. = 112320

---

18. **Answer**   A. 10.8 psig
                   B. 25.5 psia
                   C. 1558 psfg
                   D. 3672 psfa

   A. $$\frac{25 \text{ ft.}}{2.31 \text{ ft./psi}} = 10.8 \text{ psig}$$

   B. 10.8 psig + 14.7 psi (atmospheric) = 25.5 psia

   C. 10.8 psig x 144 sq.in./sq.ft. = 1558 psfg

   D. 25.5 psia x 144 sq.in./sq.ft. = 3672 psfa

---

19. **Answer**   A. 3.2 ft.
                   B. 368.6 tons additional needed

   A. Change tons to pounds; solve for psi pressure of box on water; convert to ft. That's how deep it will sink. When the pressure of the box on the water is equal to the pressure of the water up on the box, it will stop sinking.

   80 tons x 2000 lb./ton = 160,000 lb.
   This entire weight is resting on its bottom surface.

   Bottom surface = 40 ft. x 20 ft. = 800 sq.ft.

   That's $\frac{160,000 \text{ lb.}}{800 \text{ sq.ft.}}$ = 200 lb./sq.ft. = 1.39 psi

   1.39 psi x 2.31 = 3.21 ft. (pressure of box on water - in ft. of water)

   At this depth underwater, the water pressure pushing upward on the box will be the same. It will sink no further.

B. To sink it to the bottom, need enough weight to sink it to 18 ft. depth. Make a ratio:

$$\frac{80 \text{ tons}}{3.21 \text{ ft. depth}} = \frac{x \text{ tons}}{18 \text{ ft. depth}}$$

x = 448.6 tons total needed to sink it to 18 ft. depth

448.6 tons wt. needed
-    80 tons wt. of box
368.6 tons additional needed

---

20. **Answer** = 8% will show above surface.

An object will rest in water with its top just barely submerged, if its specific gravity is the same as that of water.

The volume that remains above water in an object of lesser specific gravity, is equivalent to the ratio of the two specific gravities.

$$\frac{57.2 \text{ lb./cu.ft. (iceberg)}}{62.4 \text{ lb./cu.ft. (water)}} = .92 = 92\% \text{ submerged}$$

8% above water

---

21. **Answer** = A. Water must be 3.6 ft. deep above ground to float the house.
    B. Will leak before it floats.

A. Find pressure of house on its bottom surface; convert to ft. for depth of water needed to suspend house.

House bottom surface area = 35 ft. x 60 ft. x 144 sq.in./sq.ft. = 302400 sq.in.

$$\frac{1,250,000 \text{ lb. (wt. of house)}}{302,400 \text{ sq. in}} = 4.13 \text{ psi}$$

This is equivalent to 9.6 ft. of water.

With this depth of groundwater, the house will start to float. But the house is only 6 ft. below grade, so there would have to be a 3.6 ft. deep flood in the yard to float the house.

B. To determine leaking point, convert pressure resistance of paint to equivalent water pressure in feet.

Resistance of paint = 96 inch column of oil (SG = .8)

96 inches = 8 ft.

.433 lb./ft./sq.in. (wt. of water) x .8 (SG oil) = .346 lb./ft./sq.in. oil

.346 lb./ft./sq.in. oil x 8 ft. column = 2.77 psi oil pressure = paint resistance.

The paint will hold up to 2.77 psi pressure from the groundwater.

The house will float at 4.13 psi pressure from the groundwater.

The house will leak before it floats.

CHAPTER 5

BERNOULLI'S THEOREM

STATIC SYSTEMS

From the preceding chapter, we have seen that when a vertical tube, open at the top, is installed onto a tank of water, the water will rise in the tube to the water level in the tank. This tube is called a PIEZOMETER. The water level to which the water rises in a piezometer is the PIEZOMETRIC SURFACE.

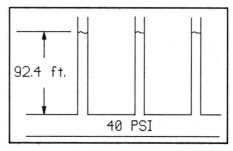

Now consider a pipeline under a constant source of pressure, with a closed valve at the end (static system). Several piezometers are installed onto the pipe. The water will rise in each of the piezometers to a height which is equivalent to the pressure in the pipe, and it will be at the same level all along the pipeline. If the pipeline pressure is 40 psi, the water will rise to 40 psi x 2.31 ft./psi: 92.4 ft. in each piezometer. It will not pour out the top. The pressure in the pipe is equivalent to 92.4 ft. of head. Once the water in each piezometer reaches 92.4 ft., its weight pushing down upon the water in the pipe is equal to the pipe's pressurized water pushing upward, and it stops rising. The width of a piezometer is of no significance. The water will rise to the same level no matter how wide it is.

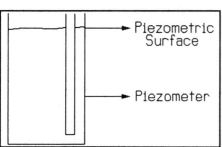

The Piezometric surface registers the level of pressure energy in the pipeline, and this level is designated all along the pipe length by an imaginary line called the HYDRAULIC GRADE LINE.

THE HYDRAULIC GRADE LINE IS THE LINE THAT CONNECTS THE PIEZOMETRIC SURFACES ALONG A PIPELINE.

Now, to complete the picture, add the pressure source, the full tank of water, with its own piezometer attached. The pressure in our pipeline originated directly from the depth of water in this tank, and each of the piezometers along the pipeline are registering that pressure. Note that the piezometer on the tank also registers the level of the water surface in the same way. In the diagram below, the valve on the end is closed. This is a static water system.

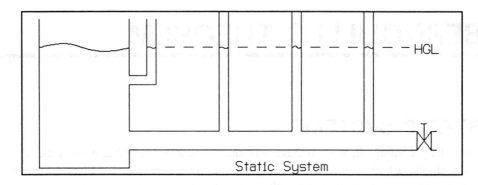

Static System

IN A STATIC WATER SYSTEM, THE HYDRAULIC GRADE LINE IS ALWAYS HORIZONTAL.

The pressure does not change along the length of that pipeline, because the water is not moving. There are no pressure losses, and the water will rise to the same height on a piezometer from any position along the pipe.

If the pipe itself is not horizontal, but curves up and down, the Hydraulic Grade Line will still remain horizontal, as long as the system is static. The pipeline pressure will be lower where the pipe is elevated, and higher where the pipe is at a lower level.

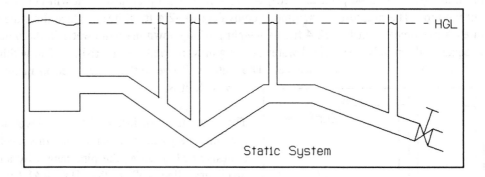

Static System

The concept of a piezometric surface is not only demonstrated in man-made pipe systems. An excellent example occurs in nature, with groundwater. The diagram shows several wells. The one in the middle is drilled into the aquifer below the water table. The well itself acts as a piezometer. Water enters it, but does not rise above the water table because this groundwater source originates from surface percolation into the ground, and this aquifer is not under pressure. The other two wells are drilled lower, into the artesian aquifer (separated from the drinking water aquifer by impermeable rock), holding pressurized water whose source is from rainfall and lake water percolation

originating at a higher elevation
(on the other side of the
mountains, in this case). These
wells, acting also as piezometers,
allow the water level to rise in
them, registering pressure from
below.   Note that one flows
freely at the ground surface
because the ground elevation is
below the level of pressure from

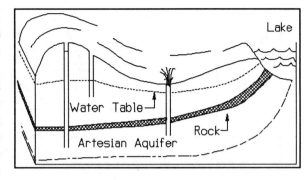

the aquifer.   The artesian water in the other does not rise all the way to the surface
because the ground level is above the level of the artesian source water.

DYNAMIC SYSTEMS - HEAD LOSS

In a Dynamic Water System, where water is flowing, the piezometric surface varies.
The Hydraulic Grade Line slopes downward along the length of pipe, showing a loss of
pressure energy.   The difference between the level of the Hydraulic Grade Line in the
static system, and that in the dynamic system, at any one point, is the feet of pressure
lost because of friction in the pipe.   This is referred to as HEAD LOSS, and is an
important factor.   A water utility must account for these pressure losses in order to
provide adequate pressure service throughout the system, when determining the water
level in the municipal water tower.

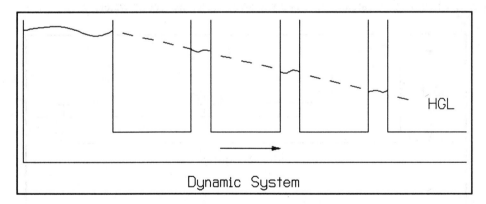

Obviously, it is not practical to install piezometers on water towers, or on pipelines.
In some cases, they would have to be hundreds of feet high.   Instead we use pressure
gages, which record pressure in feet of water, or in psi.

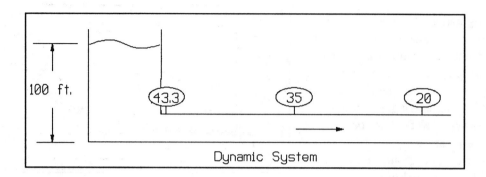

Dynamic System

TYPES OF HEAD

We know that pressure at a given point originates from the height, or depth of water above it. It is this pressure, or head, which gives the water energy, and causes it to flow. Let us consider now, the components of pressure.

Pressure Head

This is the pressure which is directly due to the depth of the water. This is familiar. If a tank of water is 80 ft. deep, the head is 80 ft. at the bottom of the tank and the psi pressure is 80 ft./2.31 ft./psi = 36.4 psi. The same pressure registers at all points along a horizontal connected pipeline, as long as the system is static. If it is a dynamic system and water is flowing, the pressure will be 34.6 psi at the bottom of the tank, and will decrease over the length of the pipeline due to friction losses in the pipe. Pressure Head, when used in calculations, is always registered as feet of water.

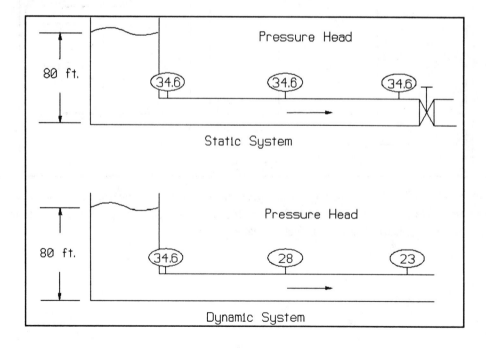

Elevation Head

If our tank of water (80 ft. deep) were up on a hill, higher in elevation than the point of measurement, the pressure in the static system would still remain level with the water surface. (HGL is always horizontal in a static system). From our point of measurement at the lower elevation, a pressure gage reading would include the depth of the water, and the elevation of the tank. This pressure due to elevation, registers on the gage and is called ELEVATION HEAD.

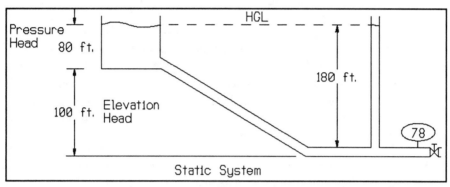

Water storage tanks are often placed at high elevations in order to take advantage of Elevation Head. Some large city water utilities have chosen to draw their raw water from more distant locations, rather than a closer source, because of the available Elevation Head to provide added pressure. Most cities have elevated water storage towers, which also employ Elevation Head to provide adequate pressure to the community.

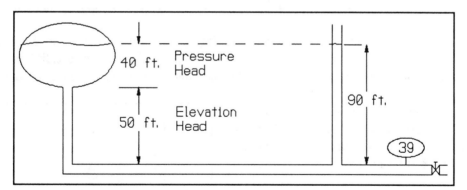

Pressure for community water use can be provided adequately by Pressure Head Alone (standpipe), or by a combination of Pressure Head and Elevation Head (elevated tower).

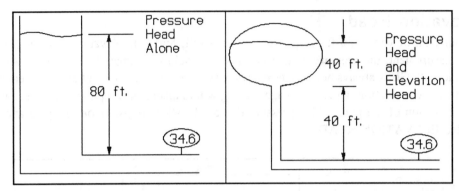

When considering Elevation Head for calculations, a point of reference, or DATUM, is always chosen, from which to measure. It can be the gage at the tank bottom, or a point on the pipeline, but the datum can also be at some unconnected point beneath the system. Pressure Head and Elevation Head can still be determined from this point, even though in this case the Elevation Head is just potential energy that is being measured, for no pipeline extends down to the datum.

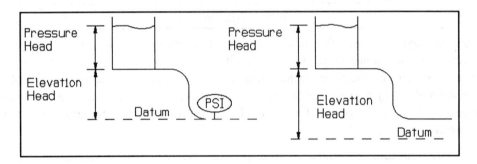

Perhaps the most practical reason for considering Elevation Head as a separate entity, is for comparison of two points along a pipeline. In determining the pressure differences between these two points, the pressure reading at Point A will be equivalent to the depth of the water in the tank (Pressure Head). The pressure reading at Point B will be equivalent to the depth of the water (Pressure Head) plus the vertical distance from Point A to Point B (Elevation Head).

Elevation Head, when used in calculations, is always registered as feet of water. It is often symbolized as Z

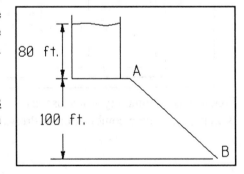

Velocity Head

There is one more energy component in a water system: Velocity Head. Moving water, just as a rolling vehicle, possesses some energy because of its motion. For instance, it takes more energy to push a car from a standstill into motion, than it does to keep it in motion. The car has some energy because it is already moving. The same applies to water which has kinetic energy because of its velocity - Velocity Head. It is the distance which the water can move due to velocity energy. Like Pressure Head and Elevation Head, Velocity Head is also expressed in feet of water. The formula for Velocity Head calculation is:

$$\text{Velocity Head} = \frac{\text{Velocity}^2}{2 \times \text{gravity}} = \frac{V^2}{2g} = \frac{V^2}{64.4}$$

Derivation:

It is derived from basic physics. When force is applied to an object, it will travel a distance, Y, which is equal to half the acceleration of the object multiplied by the square of the time involved.

$$Y = \tfrac{1}{2} a\, t^2 \quad \text{This distance is the Velocity Head.}$$

In our case, the acceleration is the force of gravity, and t can be converted to V/g (Vel. = accel. x time), (V = gt), (t = V/g)

so: $Y = \tfrac{1}{2} a\, t^2$

$Y = \tfrac{1}{2} g \times V^2/g^2$

$Y = \tfrac{1}{2} V^2/g$

$Y = V^2/2g$

In a pressurized pipe system, Velocity Head is usually the smallest of the three energy components. Normal water velocities range from 2 to 10 ft./sec. At a water velocity of 5 ft./sec., the Velocity Head would be .39 ft. Compared to Pressure Head and Elevation Head, this can be insignificant. Calculations for it are sometimes omitted. In certain instances, however, it can be very important. Where water exits freely from the end of a pipe, there is no Pressure Head left, but the water is still moving. The energy keeping it going is its Velocity Head.

Where the flow is in an open channel: an aqueduct, stream, or partially filled sewer pipe, the water is not under pressure. It has only the downward slope of the channel to maintain velocity, and Velocity Head is the energy involved.

The most useful function of Velocity Head calculations in pressurized pipe systems is its potential as a conversion unit for flow calculation. We find that Pressure Head and Elevation Head are registered on a pressure gage. Velocity Head is not. Recordable water pressure is transferred directly to the gage, because

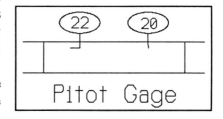

pressure in the pipe registers upward into the gage, as well as downward and to each side (Pascal's Principle). However, Velocity Head only registers forward, in the direction that the water is moving, and will not push upward into a pressure gage. A PITOT GAGE, a flow measuring device, is used to capture Velocity Head, and compare it against Pressure Head. The device may have different configurations, but is basically composed of two gages installed as a unit. One of the gages has a bent end which is directed backward into the flow. This one will record the pressure of the water (Pressure Head), plus the Velocity Head. The other gage extends straight into the pipe and registers only water pressure (Pressure Head). Converting psi readings to feet of water, then subtracting one from the other, gives us a numerical value for Velocity Head, in feet. This can now be converted to the velocity of the water with the formula $Vh = V^2/2g$. Knowing the velocity in ft./sec., and the pipe diameter, flow can easily be calculated with the Equation of Continuity ($Q=AV$). For practical use in field operations, flow dimensions can be set up on charts, based on these differential pressure readings (see end of chapter).

Using this method, differential pressure readings will enable us to calculate flow. The real importance of Velocity Head in a system under pressure, is not its magnitude as an energy component, but its capacity to become the connecting link to flow calculations.

Note that when dealing with Velocity Head calculations, we are always considering a dynamic water system. There is no Velocity Head in a static system - the water is not moving.

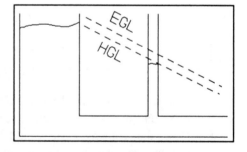

Keep in mind also that if in the system pipe diameter remains the same from point to point, then the velocity is the same at both points, and the Velocity Head is the same at both points. Velocity Head only changes from one point to another on a pipeline if the diameter of the pipe changes.

Let us refer back to the Hydraulic Grade Line as a designation of the total energy in the system. In a dynamic system, when Velocity Head is included, another grade line appears, which is called the Energy Grade Line. It is drawn just slightly above, and parallel to, the Hydraulic Grade Line. The difference between the two is that the Energy Grade Line includes Velocity Head.

TOTAL HEAD

The total energy at any one point in a hydraulic system is the sum of the Pressure Head, Elevation Head, and the Velocity Head. The summation of these energies is expressed with a formula:

Total Head = Pressure Head + Velocity Head + Elevation Head

Th = Ph + Vh + Z

All the heads must be registered in <u>feet of water</u> in order to calculate.

HEAD LOSS

When considering water traveling in a dynamic system, some pressure is lost along the length of the pipe because of friction encountered in rubbing against the pipe wall. Turbulence occurs, and energy is lost as head which then dissipates to the surroundings. The amount of loss differs according to the system length, diameter, flow, and interior roughness of the pipe. This is Friction Head Loss. It cannot be recovered, and must be calculated in by the utility when considering minimum allowable system pressures to be maintained.

Friction Head Loss is the reason for the downward slope of the Hydraulic Grade Line in a dynamic water system.

STEADY STATE FLOW

At this point, it should be emphasized that most engineering hydraulic calculations deal with a system in Steady State Flow. This means that the amount of flow crossing a given point does not change with time. Imagine our picture of a full reservoir feeding into our system and providing pressure in the pipes. For our calculations, we will assume that as the community is using this water, the water level in the tower is not dropping. If it did, then the pressure in the pipes would be constantly decreasing due to a drop in pressure head, and the flow would be constantly decreasing because there is less and less pressure pushing it through. This presents a situation which is too complicated to deal with for standard flow and pressure calculations. Therefore, we must assume that the flow and pressure source are constant. We could picture the tower constantly being filled as the system uses water, so that the flow across any one point is always the same - STEADY STATE FLOW. We understand, however, that in actual water systems, as water is being used, unsteady state flow does exist, and the utility must calculate minimum acceptable water levels.

One important fact becomes apparent in a steady state flow situation. THE FLOW THAT ENTERS THE SYSTEM, EXITS THE SYSTEM. As long as there are no leaks or lateral exits diverting the water elsewhere, it will not disappear or bunch up inside the pipe (water is incompressible); what goes in must come out. If the pipe diameter changes, the velocity will change, but the flow stays the same (Q=AV). In a dynamic system, friction losses will occur, and pressure readings will drop along the length of the pipeline, but the flow remains constant. If head losses are suddenly increased, this will cause pressure to drop, and less flow will enter the entire pipe system; but once it has entered, it must exit at the same flow rate.

BERNOULLI'S THEOREM

It was Daniel Bernoulli, a Swiss mathematician and physicist in the 1700's, who first developed the calculation for the total energy relationship from point to point in a steady state fluid system.

Bernoulli's Energy Equation holds that, ignoring friction losses, from point to point in a fluid system employing steady state flow, the total energy is the same at every point in the path of flow, and these energies are composed of Pressure Head, Velocity Head, and Elevation Head. If there is a reduction in the energy of any one form, there must be an equivalent increase in another.

Our total Head formula is now applied to a system in which we are considering two points. The formula becomes:

<u>Point 1</u>   <u>Point 2</u>

Pressure Head + Velocity Head + Elevation Head = Pressure Head + Velocity Head + Elevation Head

$$Ph_1 + Vh_1 + Z_1 \quad = \quad Ph_2 + Vh_2 + Z_2$$

A typical application is that of a pipeline in which the elevations of the two points differ. Drawing a datum horizontally through Point 2 (it is best to draw the datum through the point of lowest elevation, to avoid the use of negative numbers) we note the pressure at Point 2 is higher than that at Point 1, but the elevation of Point 1 is higher, by exactly the same amount. Inserting the numbers into Bernoulli's formula, they produce an equality. The increased Pressure Head that

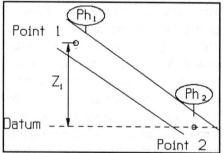

registers on the gage at Point 2, has resulted from the elevation of Point 1. Inspecting the diagram, we can see that this is obvious. Note that since the pipe diameter is the same throughout the system, the Velocity Head at both points is the same.

Taking another example, that of a pipe whose diameter differs from point to point. Point 1 lies in the pipe with the narrower diameter, Point 2 in the wider pipe. Referring back to the Equation of Continuity (Q=AV), in a steady state system, if the cross sectional area increases, the velocity of the water will decrease. Therefore, the Velocity Head at this point will also decrease. Now, Bernoulli's formula states that the total

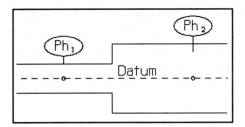

energy must remain constant from point to point. That means that one of the other forms of energy must increase at Point 2. Viewing the picture, we can see that it is not the elevation head. This is a horizontal pipe. It is the Pressure Head that increases, and by the amount that the Velocity Head decreased from Point 1. When velocity decreases, pressure energy increases.

Note that the opposite would happen if the pipe had decreased in diameter. Velocity increases, and pressure decreases. Utility personnel take advantage of this phenomenon when using the Venturi Meter, another flow measuring device. The structure of a Venturi is similar to a pipe which has been squeezed in the middle, creating a narrow diameter at the constricted section. A pressure gage is inserted at this point, and another one just upstream from the constriction. The gage at the constricted throat will read a

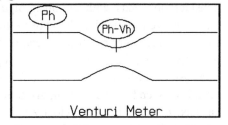

slightly lower pressure than the one upstream, and the difference, when the readings are converted to feet, will be the difference in the Velocity Heads between the two sections. At normal water pipe flows, Velocity Head at the wider section often is assumed negligible, and so the pressure differential can be equated to the Velocity Head at the constriction. This is easily convertible to a value for velocity (Vh = $V^2/2g$); then using the throat diameter and Q=AV, we can obtain a value for flow.

Be careful to note that in these situations which produce a pressure differential due to a velocity change, that it is the Velocity <u>Head</u> that is the value by which the pressure changes from point to point (not the velocity).

It should be emphasized that these internal energy changes within a pipe system, are

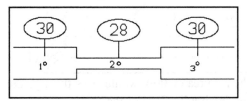

results of structural configuration, and can be reversed, by reverting back to the original structure. In this diagram, as in the Venturi Meter, the pipe constricts, and the pressure drops. The constricted section then enlarges to the original diameter, and the pressure at Point 1 is returned at Point 3.

We have now examined Bernoulli's Formula as stating the theoretical equivalence of energies from point to point in a steady state flow system. However, we have not accounted for real systems, in which there are always friction head losses.

How can we state that the summation of energy from point to point is the same when we know that in actual working systems there is always some pressure energy lost irretrievably because of friction? We can account for it - by adding friction loss into the equation also. If the total energy at Point 1 is not equal to the total energy at Point 2 because there has been some pressure lost due to friction (energy at 2 is less), then we can make it equal by adding the Head Loss to the energy at Point 2.

The formula now becomes:

$$Ph_1 + Vh_1 + Z_1 = Ph_2 + Vh_2 + Z_2 + HL_f$$

As applied to real pipe systems, and as a mathematical formula, the energies at the two points are now equal.

Do not forget - all quantities must be entered as <u>feet of water</u>.

There is another fact that should be pointed out here: when the system under consideration ends with an open pipe from which the water flows freely. In this situation, we are viewing the entire pipe system, and all the pressure energy is reduced to zero as the water reaches the end of the pipe. The Velocity Head remains, however, and this is what keeps the water moving out of the pipe. In other words, the pressure in a system is enough to overcome friction losses so that the velocity can be maintained. At the very end of the system, friction losses have been overcome, and the pressure is reduced to zero, but the velocity and flow are maintained at steady state. The flow admitted through the pipe, is the amount that will reduce the head loss to zero at the end of the pipe.

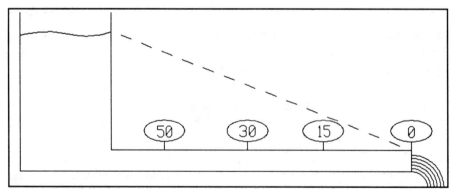

If the pressure at the source were greater, then a greater flow would pass through the pipe, and there would be a steeper drop in the Hydraulic Grade Line.

Let us consider these three diagrams below. They could depict a pipe valved at the end, or a sink faucet, or a garden hose.

In I, (Page 63) the closed valve keeps the Hydraulic Grade Line horizontal. The Energy Grade Line is superimposed upon it. They are both at the same level. There is no Velocity Head. There is no velocity. There is no flow.

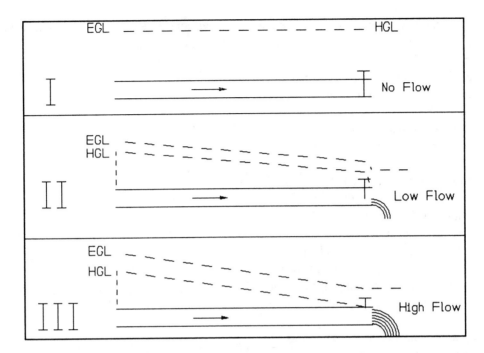

In II, the valve is cracked open slightly. A small flow is passing through the line, and out. The Hydraulic Grade Line drops only slightly along the length of the pipe. Friction losses are small because flow is low. The Energy Grade Line follows parallel to it, and slightly above, for Velocity Head (and velocity) is small. As the water encounters the nearly closed valve, two things occur. Substantial pressure energy is converted to velocity energy because the water velocity increases in passing through the narrow valve opening. Substantial friction losses also occur here. The water becomes turbulent in passing through that valve. Between these two, the water pressure on the downstream side of the valve is reduced to zero, and there is free flow out of the pipe.

In III, the valve is fully open. Great flow is allowed through. Velocity is high, and the Energy Grade Line shows significantly above the Hydraulic Grade Line. Because of the greater velocity and flow, there is greater turbulence for every foot of pipe, and the Hydraulic Grade Line decreases steeply and evenly along the length of pipe - to the end, where the pressure is zero and the Velocity Head remains to keep the water flowing.

A similar situation is encountered when there is a water main break. The system normally passes a moderate flow, with a relatively small Velocity Head, and small friction loss.

When the main ruptures, a great flood of water emerges. Flow increases enough to increase the head loss in each foot of pipe so that all the pressure is gone as the water reaches the rupture.

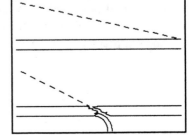

Keep in mind that in each of these diagrams, we have considered a separate system, with a different flow. It was not steady. It changed for each hydraulic condition.

One other situation should be considered at this time. When a hydraulic diagram depicts one reservoir emptying into another, and points out the differences in water levels between the two, it is meant to bring out the fact that the decreased depth of water in the second reservoir is a direct result of the head loss encountered in the pipe as the water passed from one reservoir to the other. To make this clearer, we could adjust the diagram by extending the pipeline further to the right, and narrowing in the width of the second reservoir. Now we can see that the second reservoir is nothing more than a fat piezometer, registering the pressure at that point in the pipeline.

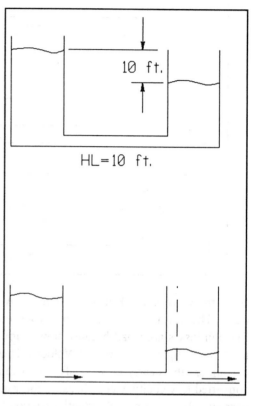

If a diagram like the one described above were to be taken literally as drawn, and water were flowing from one to the other, the flow would gradually decrease as the second reservoir filled. The head loss would also gradually decrease, because the flow was decreasing. Eventually both flow and head loss are zero, and the surface levels in the two reservoirs are the same. But this is not a steady state flow situation. It is real, but it is not steady state, and we must maintain steady state flow for calculation.

In reference to the use of Bernoulli's formula for solution of hydraulic problems - for simple situations, such as two points which differ in elevation alone, a quick view of the diagram enables us to mentally calculate for the unknown. The solution would work out just as effectively if all quantities were inserted into Bernoulli's formula, but this is not always necessary. In more complex instances, however, such as a pipe system which changes in elevation, and pipe diameter, and involves some friction loss, the usefulness of having a well defined formula into which values can be inserted is obvious, in order to avoid mistakes. In addition, as we have seen with pressure differential readings, it is Bernoulli's formula which, including Velocity Head as an energy component, allows a change in pressure to be converted to a flow quantity.

PROBLEMS

1. A flow of 2000 gpm takes place in a 12 inch pipe. Calculate the velocity head.

2. If water is flowing in a horizontal line at ground level from a 250 ft. elevated tank which holds water 20 ft. deep, at a velocity of 3.5 ft./sec., what are the:
   A. Pressure Head
   B. Velocity Head
   C. Elevation Head
   D. Total Head

3. A fluid with a Specific Gravity of 1.2 flows at 5.08 ml./sec. in a 4 inch line under a pressure of 100 psi. What is the:
   A. Pressure Head
   B. Velocity Head in this line, in feet of water

4. What is the Velocity Head in a 6 inch pipe connected to a 1 ft. pipe, if the flow in the larger pipe is 1.42 cfs?

5. A 14 inch diameter horizontal pipe under a pressure of 120 psi has a water velocity of 20 ft./sec. The pipe then constricts to a 6 inch diameter. Assuming no friction head loss, what is the pressure in the 6 inch diameter section (psi)?

6. The Total Head in a six inch diameter pipe at a point 13.5 ft. above a datum is 19.3 ft. If the pressure at that point is 312 lb./sq.ft., what is the flow (cfs)?

7. An 8 inch diameter pipe reduces to a 4 inch diameter pipe. Pressure in the 8 inch pipe is 50 psi. In the 4 inch pipe the pressure is 10 psi. Find the flow. (assume no friction loss)

8. Assume negligible head loss.
   A. Calculate the velocity in a 6 inch diameter pipe discharging 1.4 cfs to atmosphere.
   B. Calculate the velocity of this discharge through a 2.5 inch diameter nozzle at the end of the pipe.
   C. Calculate the pressure intensity (psi) in the 6 inch pipe just upstream from the nozzle.

9. An industrial water treatment process employs multi-media pressure filters for turbidity removal. Process water is pumped in an upflow mode through the filters, which have a media depth of 4 ft. The pressure loss created by turbulence and friction in passing through the media is 8 ft. If the gage reading on the 6 inch diameter influent line is 65 psi, what is the gage reading on the 6 inch diameter effluent line?

10. Water is carried from a tank holding a 25 ft. depth, through a 12 inch diameter pipe. The water flows free at 25 cfs at the end of the pipe. What is the head loss due to friction.

11. You are watering your garden with a 1 inch diameter hose. Water pressure at the faucet is 50 psi. Head loss through the hose is 4 ft. What is the pressure of the water which is being discharged from the end of the hose?

12. A ground level storage reservoir registers a pressure of 13 psi at the bottom. It services a community 100 ft. below the tank in elevation. If the total head is 131 ft., what is the velocity of the water in the pipe (assume no head loss)?

13. A 6 inch diameter pipe is slanted upward and discharges water into the atmosphere. The flow rate is 1.963 cfs and the pressure at a point in the pipe 6 ft. below the discharge end is 33.5 psi. Calculate the total head loss between that point and the discharge end.

14. A high service pump will lift water 90 ft. into an elevated tank. The gage on the pump discharge line reads 45 psi. What is the head loss due to friction (ft.)?

15. A pump is moving water from reservoir #1, which is 40 ft. in diameter and full at 20 ft. water depth. The bottom of the reservoir is 100 ft. above the centerline of the pump. The water is pumped to reservoir #2 whose top is 200 ft. above the pump centerline. Both are cylindrical tanks of the same size.
    A. How much head must the pump provide (ft.) to fill reservoir #2 (negligible head loss).
    B. How many gallons will reservoir #2 hold when full?

16. If the water pressure lost in traveling from our storage tank, which stands 50 ft. off the ground, to our new housing development is 43.3 psi how many feet of water in the tank should be provided, in order to service the housing development with a minimum 35 psi pressure?

17. At a point in a 72 inch pipe the pressure is 60 psi, and the velocity is 2 ft./sec. This pipe reduces to 36 inches, and at a point in this line, 10 ft. above the other point, the pressure reduces to 40 psi. What is the head loss from point one to point two?

18. At a point in an 8 inch diameter pipe which has an elevation of 690 ft., the pressure is 48 psi and the velocity head is .63 ft. If the headloss to a point downstream is 4.3 inches in a 10 inch diameter pipe at an elevation of 683 ft. what is the pressure in the 10 inch pipe?

19. A 4 inch diameter pitot gage attached to a hydrant registered a pressure differential of 4 psi in the gage. If the head loss caused by the gage restricted the velocity by 10%, what was the flow?

20. A pump takes water through an 8 inch suction pipe and delivers it to a 6 inch discharge pipe in which the velocity is 8 ft./sec. In the suction pipe the pressure is -6 psi. In the discharge pipe, 8 ft. above the pump, it is 60 psi.
    A. What is the pumped flow?
    B. Assuming neglig. HL, how many feet of head is the pump providing?

21. A 6 inch diameter pipe directs a jet of water vertically up into the air. If the rate of flow is 2000 gpm, how high will it go?

22. A fire pump delivers water through a 6 inch main to a hydrant. A 3 inch diameter hose is connected to the hydrant. It terminates in a 1 inch diameter nozzle. The nozzle is 10 ft. above the hydrant and 60 ft. above the pump. If friction losses are 28 ft. from pump to base of nozzle, neglecting air resistance, what gage pressure at the pump discharge is necessary to throw a stream of water 80 ft. vertically above the nozzle?

23. Water is siphoned from a tank at the rate of 20 cfs. The flowing end of the siphon is 214 ft. below the water surface. There is 1.5 ft. of headloss from tank to summit of siphon, and 1 Velocity Head is lost in friction from summit of siphon to end of siphon. The summit is 5 ft. above the water surface.
    A. Find the pipe diameter.
    B. What is the pressure at the summit?

SOLUTIONS

1. **Answer**   .5 ft.

   Change gpm to cfs; solve for Velocity; solve for Velocity Head.

   2000 gpm x 1440 min./day = 2880000 gpd = 2.88 MGD

   2.88 MGD x 1.55 cfs/MGD = 4.46 cfs.

   $$Q = A \quad V$$

   $$4.46 = .785 \times 1^2 \, V$$

   $$5.67 \text{ ft./sec.} = V$$

   $$Vh = \frac{V^2}{2g} = \frac{5.67^2}{64.4}$$

   $$Vh = .5 \text{ ft.}$$

---

2. **Answer**   A. 20 ft.
                 B. .19 ft.
                 C. 250 ft.
                 D. 270.19 ft.

   Draw Diagram. Solve for Velocity Head; add heads.

   Ph = 20 ft = depth of water in the tank.

   $$Vh = \frac{V^2}{2g} = \frac{3.5^2}{64.4} = .19 \text{ ft.}$$

   $$\cancel{Z} = 250 \text{ ft.}$$

   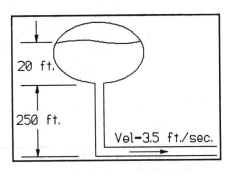

   Total Head = Hp + Vh + $\cancel{Z}$

   Total Head = 20 + .19 + 250

   Total Head = 270.19 ft.

3. **Answer**   A. 192.5 ft.
                B. 6.2×10^8 ft.

    A. Change pressure to feet of water; adjust for specific gravity.

        100 psi x 2.31 ft./psi = 231 ft. (If this liquid were water, this is Ph.)

        But SG = 1.2   so $\dfrac{231}{1.2}$= 192.5 ft.

    B. Change ml./sec. to cfs; solve for velocity; solve for Vh.

        $\dfrac{5.08 \text{ ml./sec.}}{3785 \text{ ml./gal.}}$ = .0013 gal./sec.

        $\dfrac{.0013 \text{ gal./sec.}}{7.48 \text{ gal./cu.ft.}}$ = .00018 cfs

        Q = A   V

        .00018 = .785 x $.33^2$ V

        .002 ft./sec. = V

        Vh = $\dfrac{V^2}{2g}$ = $\dfrac{.002^2}{64.4}$ = 6.2×10^8

---

4. **Answer** = .81 ft.

Draw Diagram. Flow is steady, and runs through both pipes arranged in series. Solve for velocity, then Vh.

        Q = A   V

        1.42 = .785 x $.5^2$ V

        7.24 ft./sec. = V

        Vh = $\dfrac{V^2}{2g}$ = $\dfrac{7.24^2}{64.4}$ = .81 ft.

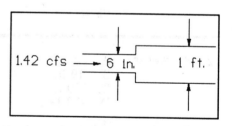

---

5. **Answer**   42.9 psi

Draw Diagram. Solve for flow; solve for Vel. in 6 inch pipe; solve for Vh in both; change psi to feet of water; using Bernoulli's formula, solve for Ph in 6 inch pipe.

        Q = A   V
        Q = .785 x 1.167^2 x 20
        Q = 21.4 cfs

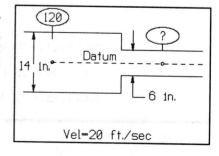

$$Q = A \quad V$$
$$21.4 = .785 \times .5^2 \, V$$
$$108.9 \text{ ft./sec.} = V$$

$$Vh = \frac{V^2}{2g}$$

$$Vh = \frac{20^2}{64.4}$$

$$Vh = 6.2 \text{ ft.}$$

$$Vh = \frac{V^2}{2g}$$

$$Vh = \frac{108.9^2}{64.4}$$

$$Vh = 184.3 \text{ ft.}$$

$$Ph_1 + Vh_1 + Z_1 = Ph_2 + Vh_2 + Z_2$$

$$277.2 + 6.2 + 0 = Ph + 184.3 + 0$$

$$99.13 \text{ ft.} = Ph$$
$$42.9 \text{ psi} = \text{pressure}$$

---

6. **Answer**   1.4 cfs

Draw Diagram.   Change pressure to feet of water; using Bernoulli's, solve for Vh, then velocity, then flow.

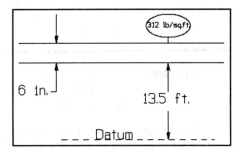

$$P = W \times H$$

312 lb./sq.ft. = 62.4 lb./cu.ft. x ft.

5 ft. = H

$$Th = Ph + Vh + Z$$
$$19.3 \text{ ft.} = 5 \text{ ft.} + Vh + 13.5 \text{ ft.}$$
$$.8 \text{ ft.} = Vh$$

$$Vh = \frac{V^2}{2g}$$

$$.8 = \frac{V^2}{64.4}$$

$$7.2 \text{ ft./sec.} = V$$

$$Q = A \quad V$$
$$Q = .785 \times .5^2 \times 7.2$$
$$Q = 1.4 \text{ cfs}$$

7. **Answer**   7 cfs

Draw Diagram. Change pressure to feet of water; setting up Bernoulli's formula it appears that not enough information is given; both Vh's are unknown; however, a relationship can be established between the velocities of the two pipes; using different assumed flows to solve for velocity, it is found

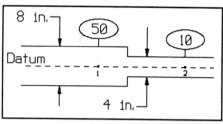

that the velocity in the smaller pipe will always be four times the velocity in the larger pipe. Obtain velocity; then solve for flow.

50 psi x 2.31 ft./psi = 115.5 ft.

10 psi x 2.31 ft./psi = 23.1 ft.

$Ph_1 + Vh_1 + Z_1 = Ph_2 + Vh_2 + Z_2$

$115.5 + \dfrac{V^2}{64.4} + 0 = 23.1 + \dfrac{(4V)^2}{64.4} + 0$

20 ft./sec. = V

79.6 ft./sec. = 4V

$Q = A \quad V$

$Q = .785 \times .67^2 \times 20$

$Q = 7$ cfs

---

8. **Answer**   A. 7.1 ft./sec.
                           B. 40.5 ft./sec.
                           C. 10.6 psi

Draw Diagram. Head loss is negligible because of short distance.

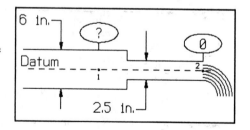

A. Solve for velocity.

$Q = A \quad V$

1.4 cfs $= .785 \times .5^2 V$

7.1 ft./sec. = V

B. Solve for velocity in nozzle.

$Q = A \quad V$

$1.4 = .785 \times .21^2 V$

40.5 ft./sec. = V

C. Use Bernoulli's formula to solve for Ph in pipe.

Nozzle:
$$Vh = \frac{V^2}{2g}$$

$$Vh = \frac{40.5^2}{64.4} = 25.4 \text{ ft.}$$

Pipe:
$$Vh = \frac{V^2}{2g}$$

$$Vh = \frac{7.1^2}{64.4} = .8 \text{ ft.}$$

$$Ph_1 + Vh_1 + Z_1 = Ph_2 + Vh_2 + Z_2$$

$$Ph_1 + .8 + 0 = 0 + 25.4 + 0$$

$$Ph = 24.6 \text{ ft.}$$

$$\text{Pressure} = 10.6 \text{ psi}$$

---

9. **Answer**   60 psi

Draw Diagram. Change pressure to feet of water; use Bernoulli's formula to solve for exit pressure head; change to psi. (Vh at both points is the same.)

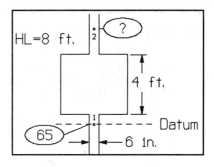

$$Ph_1 + Z_1 = Ph_2 + Z_2 + HL$$

$$150.2 + 0 = Ph_2 + 4 + 8$$

$$138.2 \text{ ft.} = Ph_2$$

$$60 \text{ psi} = \text{Pressure at effluent}$$

This problem is easy enough to solve from visual inspection once the diagram is drawn, but use of Bernoulli's formula eliminates chance of error.

---

10. **Answer**   9.2 ft.

Draw Diagram. Use Bernoulli's formula. Recognize that though all Ph is lost, some has been converted to velocity head, and some has been lost due to pipe friction. We are solving for that friction.

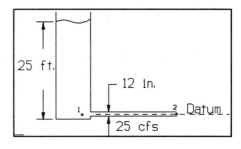

Solve for velocity; then Vh.

$$Q = A \quad V$$

$$25 = .785 \times 1^2\ V$$

$$31.9 = V$$

$$Vh = \frac{V^2}{2g}$$

$$Vh = \frac{31.9^2}{64.4} = 15.8 \text{ ft.}$$

$$Ph_1 + Vh_1 + Z_1 = Ph_2 + Vh_2 + Z_2 + HL$$

$$25 + 0 + 0 = 0 + 15.8 + 0 + HL$$

$$9.2 \text{ ft.} = HL$$

---

11. **Answer**   0 psi

Pressure discharged is flowing free, and has only velocity head.

---

12. **Answer**   8 ft./sec.

Draw Diagram. Change pressure to feet of water; use Bernoulli's formula to solve for Vh, then velocity.

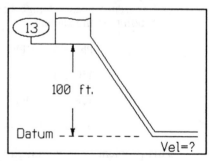

$$13 \text{ psi} \times 2.31 \text{ ft./psi} = 30 \text{ ft.}$$

$$TH = Ph + Vh + Z$$

$$131 = 30 + Vh + 100$$

$$1 \text{ ft.} = Vh$$

$$Vh = \frac{V^2}{2g}$$

$$1 = \frac{V^2}{64.4}$$

$$8 \text{ ft./sec.} = V$$

---

13. **Answer**   71.4 ft.

Draw Diagram.   Change pressure to ft. of
water; by visual inspection we can see that all
the pressure energy is lost as the water exits the
pipe. Vh remains the same at exit, and cancels
out of Bernoulli's formula, if used.

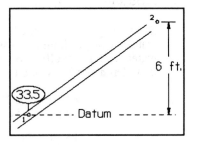

$$33.5 \text{ psi x } 2.31 \text{ ft./psi} = 77.4 \text{ ft.}$$

$$Ph_1 + Z_1 = Ph_2 + Z_2 + HL$$

$$77.4 + 0 = 0 + 6 + HL$$

$$71.4 \text{ ft.} = HL$$

---

14. **Answer**   14 ft.

Draw Diagram.   The reading on the
discharge gage registers the pressure in the
discharge pipe; in feet of water, this is the
lift. Change pressure to feet.

$$45 \text{ psi x } 2.31 \text{ ft./psi} = 104 \text{ ft.}$$

But the water is only being lifted 90 ft.
The difference is the pressure lost in pipe
because of friction.

$$104 \text{ ft.} - 90 \text{ ft.} = 14 \text{ ft. Head Loss}$$

---

15. **Answer**   A. 80 ft.
               B. 187898 gal.

Draw Diagram.

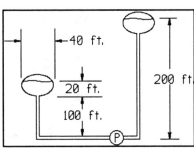

A. Water level in #1 is 120 ft. above pump.
   Pump must provide pressure from this
   level up to top of #2.

$$200 \text{ ft.} - 120 \text{ ft.} = 80 \text{ ft.}$$

B. Reservoir #2 (diam. 40 ft., depth 20 ft.)
   same size as #1. Solve for volume. Change to gallons.

$$\text{Vol.} = .785 \text{ x } d^2 \text{ x h}$$

$$\text{Vol.} = .785 \text{ x } 40^2 \text{ x } 20$$

$$\text{Vol.} = 25120 \text{ cu.ft.}$$

$$\text{Vol.} = 187898 \text{ gallons}$$

16. **Answer**   119.3 ft.

Draw Diagram. Change head loss to feet of
water; change house pressure to feet of
water; use Bernoulli's formula to solve for
tank depth (Ph_1). Velocity Heads cancel.

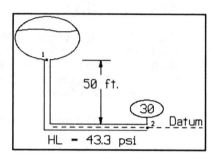

43.3 psi x 2.31 ft./psi = 100 ft.
30 psi x 2.31 ft./psi = 69.3 ft.

$Ph_1 + Vh_1 + Z_1 = Ph_2 + Vh_2 + Z_2 + HL$

$Ph_1 + 50 = 69.3 + 0 + 100$

$Ph_1 = 119.3$ ft.

---

17. **Answer**   36 ft.

Draw Diagram. Change pressures to feet;
solve for velocity at point 2; solve for Vh's
at both points; use Bernoulli's formula to
solve for Head Loss.

60 psi x 2.31 ft./psi = 138.6 ft.

40 psi x 2.31 ft./psi = 92.4 ft.

$A \quad V = A \quad V$

$.785 \times 6^2 \times 2 = .785 \times 3^2 \ V$

4 ft./sec. = V

$Vh = \dfrac{V^2}{2g}$

$Vh = \dfrac{4^2}{64.4} = .25$ ft.

$Vh = \dfrac{V^2}{2g}$

$Vh = \dfrac{2^2}{64.4} = .062$ ft.

$Ph_1 + Vh_1 + Z_1 = Ph_2 + Vh_2 + Z_2 + HL$

$138.6 + .062 + 0 = 92.4 + .25 + 10 + HL$

36 ft. = HL

18. **Answer**   51 psi

Draw Diagram. Change pressures to feet of water; solve for velocities at both points, then velocity head at point 2; use Bernoulli's formula to solve for pressure at point 2.

$$48 \text{ psi} \times 2.31 \text{ ft./psi} = 111 \text{ ft.}$$

$$\frac{4.3 \text{ inches}}{12 \text{ inches/ft.}} = .36 \text{ ft. HL}$$

$$Vh = \frac{V^2}{2g}$$

$$.63 = \frac{V^2}{64.4}$$

$$6.4 \text{ ft./sec. V}$$

$$A \ V = A \ V$$

$$.785 \times .67^2 \times 6.4 = .785 \times .83^2 \ V$$

$$4.2 \text{ ft./sec. } = V$$

$$Vh = \frac{V^2}{2g}$$

$$Vh = \frac{4.2^2}{64.4} = .27 \text{ ft.}$$

$$Ph_1 + Vh_1 + Z_1 = Ph_2 + Vh_2 + Z_2 + HL$$

$$111 + .63 + 7 = Ph_2 + .27 + 0 + .36$$

$$118 \text{ ft. } = Ph_2$$

$$51 \text{ psi } = \text{Pressure}$$

---

19. **Answer**   1.88 cfs

Pressure differential on pitot gage is the Vh. Change it to feet of water. Then to solve for flow we must solve for velocity, then flow. HL restricts vel. (or flow) by 10%. Reduce flow amount by 10%.

$$4 \text{ psi} \times 2.31 \text{ ft./psi} = 9.24 \text{ ft.}$$

$$Vh = \frac{V^2}{2g}$$

$$9.24 = \frac{V^2}{64.4}$$

$$24.4 \text{ ft./sec. } = V$$

$$Q = A \ V$$

$$Q = .785 \times .33^2 \times 24.4$$

$$Q = 2.09 \text{ cfs}$$

$$90\% \times 2.09 = 1.88 \text{ cfs}$$

---

20. **Answer**   A. 1.57 cfs
                 B. 161.2 ft.

A. Solve for flow.

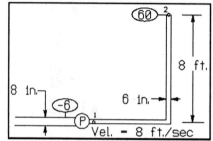

$$Q = A \ V$$

$$Q = .785 \times .5^2 \times 8$$

$$Q = 1.57 \text{ cfs}$$

B. Draw Diagram. Change pressures to feet of water. Solve for velocity heads on both suction and discharge sides. Solve for Total Head on each side. Subtract suction head from discharge head to yield head which pump is providing.

Suction Side
$$Q = A \ V$$

$$1.57 = .785 \times .67^2 \times V$$

$$4.46 \text{ ft./sec.} = V$$

$$Vh = \frac{V^2}{2g}$$

$$Vh = \frac{4.46^2}{64.4} \text{ ft.}$$

$$Vh = .3 \text{ feet}$$

Discharge Side
$$Vh = \frac{V^2}{2g}$$

$$Vh = \frac{8^2}{64.4}$$

$$Vh = 1 \text{ ft.}$$

Suction Side
$$TH = Ph + Vh + \cancel{Z}$$

$$TH = (-13.9) + .3 + 0$$

$$TH = -13.6 \text{ ft.}$$

Discharge Side - (at gage)
TH = Ph + Vh + ~~Z~~

TH = 138.6 + 1 + 8

TH = 147.6 ft.

Discharge Head - Suction Head = Pumping Head

147.6 - (-13.6) = 161.2

Pump provides 161.2 ft. of head

---

21. **Answer**   .8 ft.

Draw Diagram. The jet will travel up a distance that is equivalent to the velocity head; this is the only energy it has at this point. Change gpm to cfs; solve for velocity at exit; solve for VH.

2000 gpm x 1440 min./day = 2880000 gpd = 2.88 MGD

2.88 MGD x 1.55 cfs/MGD = 4.46 cfs

Q = A V

4.46 = .785 x $.5^2$ V

22.7 ft./sec. = V

$Vh = \dfrac{V^2}{2g}$

$Vh = \dfrac{22.7^2}{64.4} = .8$ ft.

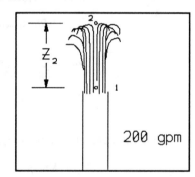

200 gpm

Bernoulli's formula can be used to prove that the elevation distance of travel is equal to Vh.

$Ph_1 + Vh_1 + Z_1 = Ph_2 + Vh_2 + Z_2 + HL$

0 + .8 + 0 = 0 + 0 + ~~Z~~ + 0

.8 ft. = Z_2

22. **Answer**   73 psi

Draw Diagram. Set point 1 at pump discharge, and point 2 at end of nozzle.

Use Bernoulli's formula to determine pressure at pump. Solve for Vh at pump first, then insert all quantities into Bernoulli's formula.

$$Vh = \frac{V^2}{2g}$$

$$80 = \frac{V^2}{64.4}$$

$$V = 71.8 \text{ ft./sec}$$

$$A \ V = A \ V$$

$$.785 \text{ x } .5^2 \text{ V} = .785 \text{ x } .083^2 \text{ x } 71.8$$

$$V = 1.98 \text{ ft./sec}$$

$$Vh = \frac{V^2}{2g} = \frac{1.98^2}{64.4}$$

$$Vh = .061 \text{ ft.}$$

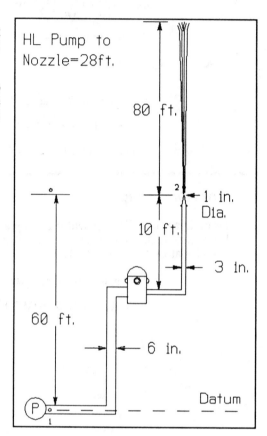

HL Pump to Nozzle=28ft.

80 ft.

2 — 1 in. Dia.

10 ft.

3 in.

60 ft.

6 in.

Datum

Velocity head at this point could be eliminated; it is negligible.

$$Ph_1 + Vh_1 + Z_1 = Ph_2 + Vh_2 + Z_2 + HL$$

$$Ph_1 + .061 + 0 = 0 + 80 + 60 + 28$$

$$Ph_1 = 168 \text{ ft.}$$

$$\text{Pressure} = 73 \text{ psi}$$

Vh at point 2 is 80 ft. (the pump is throwing water 80 ft. up)

23. **Answer**

A. 6 inch diameter

B. -6.5 ft. pressure at summit

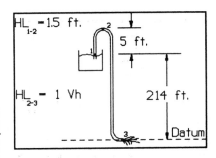

Draw Diagram.

Solve for Ph at summit (point 1 at water surface, point 2 at summit of siphon); solve for HL in downward length of siphon; this equal 1Vh; solve for velocity; then diameter.

$$Ph_1 + Z_1 = Ph_2 + Z_2 + HL$$

$$0 + 214 = Ph_2 + 219 + 1.5$$

$$Ph_2 = -6.5 \text{ ft.}$$

$$Ph_2 + Z_2 = Ph_3 + Z_3 + HL$$

$$-6.5 + 219 = 0 + 0 + HL$$

$$212.5 \text{ ft.} = HL$$

$$212.5 \text{ ft.} = 1Vh$$

$$Vh = \frac{V^2}{2g}$$

$$212.5 = \frac{V^2}{64.4}$$

$$117 \text{ ft./sec.} = V$$

$$Q = A \ V$$

$$20 = .785d^2 \times 117$$

$$.47 \text{ ft.} = d$$

$$6 \text{ inches} = d$$

CHAPTER 6

PUMPING - INTRODUCTION

PUMPING HEADS

In the last chapter we noted the components of pressure in any moving water system (Pressure Head, Velocity Head, Elevation Head), and we studied Bernoulli's energy formula. Now we will take a look at the specialized designation of heads common to pumping systems. A pump is installed to move water when the pressure is inadequate to do the job. For instance, water will not move from the treatment plant up into the elevated storage tank on its own. The High Lift pumps provide the needed pressure to lift the water to this height. In determining the amount of pressure a pump must provide to get the job done, the nomenclature of PUMP HEADS was established.

Inspect the diagram below:

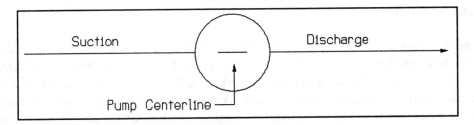

Note the SUCTION and DISCHARGE side of the pump. Water enters the pump on the suction side. Here the pressure is lower. Since the <u>function of the pump is to add pressure to the system</u>, discharge pressure will always be higher. In pump systems, measurements are taken from the point of reference to the CENTERLINE of the pump (horizontal line drawn through center of pump). Recall the distinction between:

STATIC WATER SYSTEM - water not moving

DYNAMIC WATER SYSTEM - water moving

Now let's give this pump a job of lifting water from a lower tank to a higher tank.

Static Heads

STATIC SYSTEM - <u>Pump is OFF!</u>

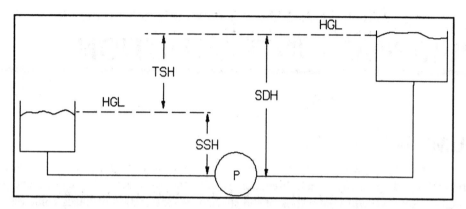

The vertical distance from the water surface on the suction side, to the pump centerline, is the STATIC SUCTION HEAD (SSH). Static Head is always a positive value. It equates to pressure already in the system - work that is provided by the system, which the pump does not have to do. If the pump were not installed in this system at all, the pressure in the system would lift the water to this same height on the right side of the diagram. The pump did not have to provide this much of the pressure.

Static Suction Head is registered on the gage at the suction side of the pump - when the pump is off (static system).

Referring back to the diagram, the vertical distance from the pump centerline to the water level on the discharge side of the pump is the STATIC DISCHARGE HEAD. This is theoretically the total amount of work done by the system. The water was actually lifted this distance. Some of this work was done by the pump; some was done by the water pressure originating from the height of the tank on the suction side. The gage on the discharge side of the pump in a static system, will register Static Discharge Head.

The vertical difference between the Static Suction Head and the Static Discharge Head is called TOTAL STATIC HEAD. This is the theoretical amount of work that the pump must do in lifting water up to the higher tank. To provide this much head is why we have purchased this pump!

Now let us consider the situation below. Pumps are frequently employed to draw water up from below the level of the pump, as in the case of a well. The configuration is such that on the suction side there is a lift, a STATIC SUCTION LIFT. The pump must have the energy to provide this lift. On the discharge side, in this case, the pump is also providing the head needed to raise the water to the tank above, the STATIC DISCHARGE HEAD. In pump systems which provide a suction lift, the TOTAL

STATIC HEAD is the sum of the Static Suction Lift and the Static Discharge Head. The pump is doing <u>all</u> the work.

STATIC SYSTEM - <u>Pump is Off!</u>

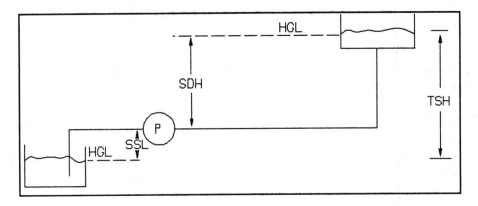

Dynamic Heads

If a pump system is to provide pressure to move water, obviously the pump must be on. This is the actual working situation. Suction and discharge heads in a working system are not static. They are dynamic heads, and they will depend upon the length and condition of the system. As water moves through a pipeline, it loses pressure because of friction encountered in transit (friction head loss). Therefore suction and discharge heads will decrease with pipe length. The Hydraulic Grade Line will slope downward toward the pump on the suction side, and downward away from the pump on the discharge side, designating that some pressure has been lost because this water is moving; the system is dynamic.

DYNAMIC SYSTEM - <u>Pump is On!</u>

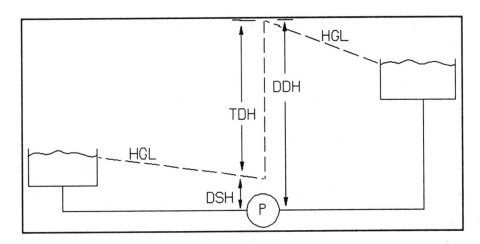

The DYNAMIC SUCTION HEAD is the vertical distance between the water level on the suction side and the centerline of the pump, <u>minus the friction headloss</u>.

The DYNAMIC DISCHARGE HEAD is the vertical distance between the pump centerline and the water level on the discharge side, <u>plus the friction headloss</u>. The pump must make up this pressure lost in transit in order to get the water up into the tank.

The Dynamic Discharge Head is the pressure that the pump discharge gage will record when the pump is on.

The TOTAL DYNAMIC HEAD is the difference between the Dynamic Suction Head and the Dynamic Discharge Head. Note that the actual work this pump has to do in lifting the water to the higher tank, is greater than that designated by static heads. The pump must make up friction losses on both suction and discharge sides.

Again, look at the case of a suction lift (below). Since pressure is lost in transit to the pump, the DYNAMIC SUCTION LIFT is the vertical distance between the water level and the centerline of the pump, <u>plus the friction headloss</u>. It will register as negative pressure (amount of vacuum) on a gage on the suction side of the pump.

The DYNAMIC DISCHARGE HEAD is the vertical distance between the pump centerline and the water level on the discharge side, <u>plus the friction headloss</u>. The TOTAL DYNAMIC HEAD in this system is the sum of the Dynamic Suction Lift and the Dynamic Discharge Head - and again, this includes the friction losses on both sides of the pump. This is the amount of pressure the pump must provide in this system, the energy needed to move the water. It is the pumping head which determines the size of pump needed for this operation.

DYNAMIC SYSTEM - <u>Pump is On!</u>

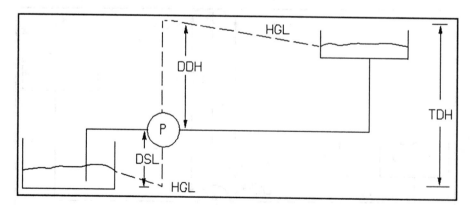

The diagrams above have pictured an actual lift situation. The pump was raising the water to a higher elevation. However, the same pumping heads can be identified in a booster pumping situation. In this case the line pressure available from the pressure source (an elevated tank, perhaps) has dropped to too low a level, and a booster pump is installed to boost pressure. The suction and discharge lines extend laterally. There does not have to be a change in elevation. The pump is installed solely to restore the pressure depleted by friction losses in the system.

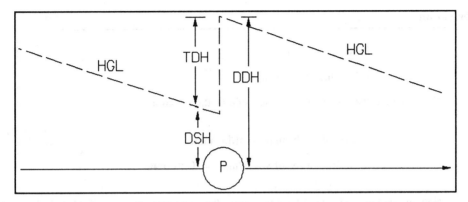

Dynamic Suction Head is the line pressure on the suction side of the booster pump. Dynamic Discharge Head is the line pressure on the discharge side of the pump. Total Dynamic Head is the pressure that the booster pump is providing.

POWER CALCULATIONS

In order to calculate the cost of operating a pump, it is necessary to equate the amount of water moved, to units of electrical energy consumed in moving that water. In order to purchase a pump, it must first be determined how much work that pump will be required to do.

We must start with the concept of <u>work</u>, the operation of a force over a specific distance. The unit of work is the <u>foot pound</u>: the amount of work required to lift a one pound object one foot off the ground (ft.lb.).

When considering work being done, it is more valuable to be able to do the work faster, and so we consider the rate at which work is being done. This is called <u>Power</u>, and is labelled as <u>foot pound/second</u>. At some point it was determined that the ideal work animal, the horse, could move 550 pounds one foot, in one second. Since large amounts of work are also to be considered, this unit became known as <u>Horsepower</u>.

550 ft.lb./sec. = 1 Horsepower

or

33,000 ft.lb./min. = 1 Horsepower

A pump also does work. It also has power. It lifts water (which has weight) a given distance, in a specific amount of time. The unit label is the same - ft.lb./min. We state an equality:

Derivation:

Power = Power

ft.lb./min. = ft.lb./min.

Now separate the components of one side of the equation.

feet of lift x lb./min. = ft.lb./min.

ft. x flow in lb. of water/min. = ft.lb./min.

Now divide both sides by 33000 ft.lb./min./hp - and also divide by 8.34 lb./gal. to change lb. of water to gal. of water.

$$\frac{\text{ft. x lb./min.}}{33000 \text{ ft.lb./min./hp. x } 8.34 \text{ lb./gal.}} = \frac{\text{ft.lb./min}}{33000 \text{ ft.lb./min./hp.}}$$

$$\frac{\text{ft. x gal./min.}}{3960} = hp$$

This condensed version is the standard formula for horsepower, derived from the amount of lift, or head, a pump will be expected to provide (in ft.) and the flow at which the pump will be operating (gpm).

This is WATER HORSEPOWER, the power necessary to move the water to this height.

$$\frac{\text{ft. x gpm}}{3960} = wph$$

But a pump does not operate alone. A motor drives the pump, and electrical energy drives the motor. Neither pump nor motor are 100% efficient. There are friction losses even within both of these units, and it will take <u>more</u> horsepower applied to the pump to get the required amount of horsepower to move the water, and even <u>more</u> horsepower applied to the motor to get the job done.

BRAKE HORSEPOWER is the horsepower applied to the pump.

MOTOR HORSEPOWER is the horsepower applied to the motor.

EFFICIENCY is the power produced by the unit, divided by the power used in operating the unit.

The Water Horsepower must be divided by the pump efficiency to obtain the Brake Horsepower. The Brake Horsepower must be divided by the motor efficiency to obtain the Motor Horsepower.

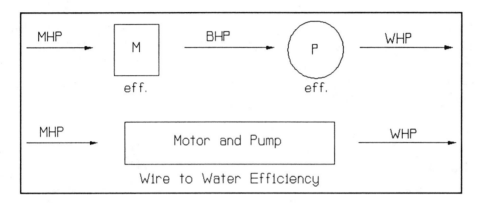

Most pumps operate at 60-85% efficiency.

Motors are usually 80-90% efficient.

Multiplying the pump efficiency by the motor efficiency will provide the combined, or total efficiency of the system, the WIRE TO WATER EFFICIENCY.

Now, with the ability to calculate the actual power necessary to apply to the motor in order to do the required work, we need to convert that power from units of Motor Horsepower, to units of electrical energy (kilowatts). There is a direct conversion:

$$1 \text{ Horsepower} = .746 \text{ Kilowatts}$$

This power demand, in Kilowatts, is sold by the hour (Kilowatt hours). Multiply then, by the number of hours the pump runs, and then by the cost per Kilowatt Hour, and we have the cost of running the pump.

In summary:

$$\frac{\text{gpm x head}}{3960} = \frac{\text{water Hp}}{\text{pump effic.}} = \frac{\text{brake Hp}}{\text{motor effic.}} = \text{motor Hp x } .746 \frac{\text{Kw}}{\text{Hp}} \text{ x hrs. x } \frac{\text{cost}}{\text{Kw.hr.}} = \$$$

WELL PROBLEMS - PUMPING PROBLEMS

This is a basic category of calculations which apply to water pumped from a groundwater aquifer, as well as to water pumped from a wastewater wet well or tank in a suction lift situation. The basic principles, terminology, and calculations apply to both.

Groundwater Well

Well Yield - the volume of water pumped from a well in a specified time period. Well yield is expressed as gallons per minute (gpm).

Drawdown - the distance the water level drops once pumping begins. The vacuum created at the end of the pump intake line brings water from the immediate area into the well, and a depression is caused in the water level in that area. Groundwater travel is restricted by the soil as it moves toward the well, and the difficulty encountered in replenishing the water taken up by the pump is registered as a lowered pumping water level. It is important to note that once drawdown exists, the pump chosen must be powerful enough to lift the water that extra distance, and still provide the needed capacity.

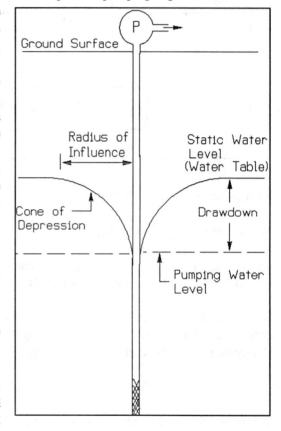

Static Water Level - the water level when the pump is off.

Pumping Water Level - the water level when the pump is on.

The drawdown is the difference, measured in feet, between the Static Water Level and the Pumping Water Level.

Safe Well Yield - the maximum pumping rate that can be achieved without increasing the drawdown. In a groundwater well, the Drawdown may easily be fifty feet or more. It may take an hour or so to stabilize this Drawdown and arrive at a permanent Pumping Water Level, but once this is achieved, the Drawdown should not be increasing over time. If the Drawdown is 20 ft. this month, it should be 20 ft. next month, and next year, regardless of whether that pump is run continuously, or in an on/off mode. If the Drawdown increases as time passes, then the pump is withdrawing more water then the aquifer can return to the well. This is called Water Mining. Eventually, the Pumping Water Level will drop below the end of the well screen, and the well will fail.

Overpumping an aquifer can have permanent effects on the land itself: sinkhole formation, and in areas proximate to the ocean, saltwater intrusion. Note that natural phenomena can also influence well supply. Drought lowers the Static Water Level, so that even with acceptable Drawdown, the Pumping Water Level may be too low in dry years for effective pumping (many small private wells run dry during times of drought). Excessive rainfall raises the Static Water Level, reversing this effect. Soil porosity and geologic structure also have great influence on the aquifer's ability to return groundwater to the well, and will limit pump size.

Radius of Influence - the distance between the pump shaft and the outermost area affected by drawdown. This becomes important in well fields with many pumps. If wells are set too close together, the Radii of Influence will overlap, increasing the drawdown in all wells. Pumps should be spaced far enough apart so that this does not happen.

Cone of Depression - the curve of the line which extends from the Pumping Water Level to the Static Water Level at the outside edge of the Radius of Influence. Engineering calculations based on the slope of this curve will determine the Radius of Influence.

Specific Capacity - pumping rate per foot of Drawdown (gpm./ft.). This is a measure of Well Yield. Given the same pumping rate, if the Specific Capacity is decreasing with time, the Drawdown is increasing.

Wet Wells

Water pumped from a tank by a pump set above the water surface exhibits the same phenomena as groundwater wells. There is a slight depression of the water surface right at the intake line (drawdown) while the pump is operating, but in this case it is minimal because there is no restriction of the water approaching the pump entrance. As the pump draws water up, it fills into that space almost immediately, so that Pumping Water Level is barely lower then Static Water Level. Most important at this location is to ascertain that the suction line is submerged far enough below the surface, so that air entrained by the active movement of the water at this section is not able to enter the pump.

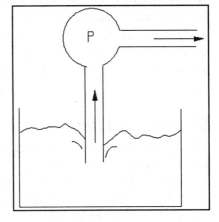

If the pump is operating at exactly the same rate as the tank is filling, and if the pump is on continuously, then the water level in the tank will remain the same. However, for water and wastewater applications, this is very rarely the case. Water enters the tank at varying flows, depending on the source, and since wet well size is limited, pumping ability must accommodate for large and small flows, and will almost never exactly match the inflow. To cope with this, variable speed pumps can be used, or several pumps are installed for single or combined operation. In some cases pumping is done in an on/off mode, as with sump pumps. Control of pump operation is in response to water level in the well. Level control devices will sense a high and low level in the well, and transmit the signal to the pumps for action. In wet well calculations, any lowering of the water level in the well is referred to as Drawdown. A common method for calculating pump performance (pumping rate) is to run a pumping test. Water is pumped out of a tank of known volume, drop in water level is recorded (drawdown), and that volume of water is converted to gallons. Divide by the time span of the pump test, and this will yield gallons per minute pumping rate. Note that if there is a simultaneous inflow to the well, this will have to be calculated also.

A water meter may be installed on the pump discharge line to monitor pumping rate, but a pumping test can be useful to check the accuracy of the meter.

WATER LEVEL SENSING AND PUMP CONTROL

Accurate sensing of high and low water levels in a wet well is imperative for proper timing of pump operation.

In groundwater well maintenance, accurate recording of Static Water Level, Pumping Water Level, and Drawdown, is just as important.

There are several different types of water level sensing devices in common use:

Mechanical Float System

In use for many years, this system is installed in tanks and wet wells to sense high and low water levels for signaling pump operation. Steel rods have attached floats which extend from the side of the tank and ride up and down on the water surface. At maximum and minimum level the system mechanically activates a transmitter which signals the pump to action. These systems are bulky, but simple in design and reliable.

Diaphragm Element

This system, also in use with tanks, operates with a small chamber filled with air. The bottom section of it is composed of a flexible diaphragm, extended outward when the water level is low. The level rises, and at a specific depth the pressure generated by the water depth pushes the diaphragm inward, compressing the volume of air inside. This increased air pressure is registered on a sensor and transmitted to the pump to turn on. This type of system is often installed into the bottom of sump pumps. The water level rising over the pump switches the pump into operation. As the water level goes down, pressure decreases, the diaphragm extends outward again, and the pump switches off.

Bubbler System

This is probably the most common system of all, and is used in high and low water level sensing in wet wells, as well as for reading water levels in groundwater wells. It requires a source of air (compressor or tire pump) which is pumped through a narrow copper tube extending to the bottom of the well or tank. With the air source turned off, the water will fill the tube to the level of the surface. Air pumped into the tube expels the water in it. As soon as all the water is evacuated from the tube, air bubbles out the bottom. A pressure gage at the top of the airline records the pressure needed to accomplish this, which, in feet, is equivalent to the water depth in the tank or well. If a pressure reading is recorded when a well pump is off (Static Water Level), and when the pump is on (Pumping Water Level) - the difference between these readings is the Drawdown. Wastewater lift stations commonly employ this system to signal high and low water levels for pump operation.

Electrical Sounder

A battery is connected to a pair of insulated wires, and lowered into the well until it touches water, which completes the circuit. An ammeter, or light shows the flow of current, and the depth to water is read from calibrations on the wires. Wastewater lift stations use this mechanism as a high water alarm, but care must be taken to keep the mechanism clean for proper operation. In a groundwater well, it may be installed into a separate stilling well on either inside or outside of pump casing to avoid interfering with the pump column.

Electronic/Pneumatic Transducers

Electronic probes are installed at known depths and sense water rising over them. The signal is transmitted to the surface and displayed. Pneumatic transducers provide an output signal equivalent to the gas pressure in the line which is equivalent to the water pressure in the well. Both of these types have been developed recently, and provide good accuracy. However, care must be taken to avoid grease accumulations on the probes.

Ultrasonic Detectors

This is based on high frequency sound waves. A transmitter is mounted in the wet well, pointed vertically downward. Sound waves are sent to the water surface; they bounce back, and the time span is measured. This is electronically converted to a milliamp signal which is proportional to the time, and to the water level. This system is in popular use in wastewater treatment, for no part of the detector touches the water.

PROBLEMS

1. Water is pumped at 500 gpm to a water storage tank on the roof of an industry. The gage on the pump discharge line reads 95 psi. The difference in elevation between the gage reading and the water level in the tank is 90 ft. What is the pressure loss (ft.) in the piping?

2. A wet well pump operating with a suction lift has a Total Dynamic Head of 150 ft. If the Dynamic Discharge Head is 130 ft., and the friction head loss on the suction side is 10 ft., what is the vertical distance between the water level on the suction side and the centerline of the pump?

3. A flow of 1840 gpm is to be pumped against a head of 40 ft. What is the water horsepower required?

4. A 12 hp pump delivers 1200 gpm. Assuming the pump is 100% efficient, what is the pressure (psi) against which the pump is operating?

5. A pump is putting out 5 whp and delivering a flow of 430 gpm. Assuming no friction head loss, what is the height to which this pump is lifting water?

6. A pump is 75% efficient, and the motor is 82% efficient. If the pump is delivering 825 gpm against a Total Dynamic Head of 95 ft.:
   A. What is the Water Horsepower?
   B. What is the Brake Horsepower?
   C. What is the Motor Horsepower?

7. If the motor is 94% efficient and the pump is 90% efficient, what horsepower will be required to deliver 200 gpm against a total pressure of 45 psi?

8. A motor draws 2238 Watts. The weekly charge for operating the pump is $1.65. If the cost per Kilowatt Hour is $.015, how many hours per week is the pump in operation?

9. A 25 hp pump is operating against a total pressure of 30 psi. If the efficiency of the pump is 88%, what is the maximum gpm delivery?

10. A pump 86% efficient delivers 590 gpm with a Brake Horsepower of 6.5 hp. What is the suction lift in feet if all other heads are equivalent to 34 ft.?

11. A 5 hp pump/motor combination is pumping 355 gpm against a Total Dynamic Head of 41 ft. If the motor is 91% efficient, what is the pump efficiency?

12. A pump is discharging 700 gpm against a head of 82 ft. The wire to water efficiency is 83% and the power rate is $.08/Kw.hr. What is the power cost for 14 hours of pumping?

13. A 300 gpm centrifugal pump is operating against 15 ft. of lift and losses on the suction side, and a discharge pressure of 50 psi. Instruments placed in the motor circuit show that it consumes 5.5 Kw.hr. in 30 minutes. What is the wire to water efficiency?

14. A water level gage on a 500 gpm well pump registers 52 psi when the pump is off. When the pump is running, this gage registers 39 psi.
     A. What is the Drawdown?
     B. What is the Specific Capacity of the well?

15. Given the following information on a well:
             Static Water Level: 400 ft.
             Casing diameter: 12 inches
             Well depth: 1200 ft.
             Pumping rate: 150 gpm
             Pumping Water Level: 500 ft.
     A. What is the Drawdown when the pump is operating?
     B. What is the Specific Capacity?

16. Checking water level in a well. When 400 gpm is being pumped the level gage pressure reading is 29.5 psi and the Drawdown is 52 ft. When the pump is shut off the water level rises to 200 ft. below ground surface. What is the pressure reading at that time (psi)?

17. An airline water level gage 200 ft. long extends from the ground surface to the bottom of a well. With the well pump off, the line is pressurized and a pressure of 55 psi is read on the gage. Well pump is then turned on, and after stabilizing, 32 psi is read.
     A. What is the depth to the water table?
     B. What is the Drawdown?

18. A tank 20 ft. in diameter and 10 ft. deep is half full. Flow entering equals 100 gpm. Pumping out 50 gpm. How long does it take to fill the tank?

19. A wastewater wet well is 10 ft. in diameter and 9 ft. deep. With no pump operating, the rise in water level is 2 ft. in 1 min. 46 sec. A pump is started and the water level begins to drop at a rate of 4 inches in 4 minutes. What is the gpm pumping rate?

20. During a 30 minute pumping test, 3680 gallons are pumped into a tank which has a diameter of 10 ft. The water level before the pumping test was 3 ft. What is the pumping rate in gpm?

21. Water is pumped into an empty 22 ft. diameter tank at the rate of 200 gpm. At the end of 4 hours of pumping, what is the gage reading at the bottom of the tank (psi)?

22. A community's 75,000 gallon elevated storage tank is one third full. The wells are pumping at a steady rate of 500 gpm. The distribution system demand is constant at 200 gpm. How much time will elapse before the tank is full (hrs.)?

23. A lift station wet well is pumped at a rate of 150 gpm for 20 minutes each hour. A constant flow of water is coming into the station at a rate of .1 cfs. What is the problem with this station, and offer a positive solution for it.

24. A wet well is 11 ft. by 11 ft. by 10 ft. 6 inches deep. The influent rate is 110 cfm, and when the pump is operating, you measure a drop in the water level of 15 inches in 4 min. 20 sec. If your pump is designed for 1200 gpm delivery, what percent of design efficiency is the pump operating at?

SOLUTIONS

1. **Answer**   130 ft.

Draw Diagram. Pump lifts water to storage tank. Discharge gage registers Dynamic Discharge Head (lift plus losses). Change pressure to feet of water; solve for head loss.

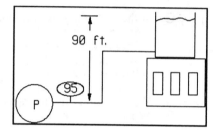

95 psi x 2.31 ft./psi = 220 ft.

DDH = Lift + Losses

220 = 90 + x

130 ft. = x

2. **Answer**   10 ft.

Draw Diagram.   Total Dynamic Head includes all the work the pump is doing (lift & losses on both suction and discharge sides). Dynamic Discharge Head = lift & losses on the discharge side. Solve for suction lift.

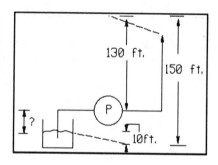

TDH = DSL + DDH

150 = (x + 10) + 130

10 ft. = x

3. **Answer**   18.6 hp

$$\frac{gpm \times hd}{3960} = whp$$

$$\frac{1840 \times 40}{3960} = whp$$

18.6 = whp

4. **Answer**   17.2 psi

   Solve for head; change to pressure.

   $$\frac{\text{gpm x hd}}{3960} = \text{whp}$$

   $$\frac{1200 \text{ x hd}}{3960} = 12$$

   Hd = 40 ft.

   $$\frac{40 \text{ ft.}}{2.31 \text{ ft./psi}} = 17.2 \text{ psi}$$

---

5. **Answer**   46 ft.

   $$\frac{\text{gpm x hd}}{3960} = \text{whp}$$

   $$\frac{430 \text{ x hd}}{3960} = 5$$

   hd = 46 ft.

---

6. **Answer**   A. 19.8 hp
                  B. 26.4 hp
                  C. 32.2 hp

   A. $$\frac{\text{gpm x hd}}{3960} = \text{whp}$$

   $$\frac{825 \text{ x } 95}{3960} = \text{whp}$$

   19.8 = whp

   B. $$\frac{\text{whp}}{\text{pump eff.}} = \text{bhp}$$

   $$\frac{19.8}{.75} = \text{bhp}$$

   26.4 = bhp

   C. $$\frac{\text{bph}}{\text{motor eff.}} = \text{mhp}$$

   $$\frac{26.4}{.82} = \text{mhp}$$

   32.2 = mhp

7. **Answer**   6.2 hp

Change pressure to feet of water; solve for motor horsepower.

$$45 \text{ psi} \times 2.31 \text{ ft./psi} = 104 \text{ ft.}$$

$$\frac{\text{gpm} \times \text{ft.}}{3960} = \text{whp}$$

$$\frac{200 \times 104}{3960} = \text{whp}$$

$$5.3 = \text{whp}$$

$$\frac{\text{whp}}{\text{pump eff.} \times \text{motor eff.}} = \text{mhp}$$

$$\frac{5.3}{.94 \times .90} = \text{mhp}$$

$$6.2 = \text{mhp}$$

---

8. **Answer**   49 hr./week

Change watts to Kilowatts; solve for hours.

$$\frac{2238 \text{ watts}}{1000 \text{ watts/kw}} = 2.238 \text{ Kw}$$

$$\text{Kw} \times \text{hr.} \times \text{cost/Kw.hr.} = \text{Total cost}$$

$$2.238 \times \text{hr.} \times .015 = 1.65$$

$$\text{hr./week} = 49$$

---

9. **Answer**   1257 gpm

"Operating against a pressure of" means "Total Dynamic Head" - the work the pump is doing. A horsepower rating given as the property of the pump is the Brake Horsepower. Solve for whp; solve for gpm.

$$\frac{\text{whp}}{\text{pump eff.}} = \text{bhp}$$

$$\frac{\text{whp}}{.88} = 25$$

$$\text{whp} = 22$$

$$\frac{\text{gpm x hd}}{3960} = \text{whp}$$

$$\frac{\text{gpm x 69.3}}{3960} = 22$$

$$\text{gpm} = 1257$$

10. **Answer**   3.6 ft.

Solve for whp (it is the power needed to move the water); solve for TDH (work the pump has to do-includes suction and discharge); solve for suction lift.

$$\frac{\text{whp}}{\text{pump eff.}} = \text{bhp}$$

$$\frac{\text{whp}}{.86} = 6.5$$

$$\text{whp} = 5.6$$

$$\frac{\text{gpm x hd}}{3960} = \text{whp}$$

$$\frac{590 \text{ x hd}}{3960} = 5.6$$

$$\text{hd} = 37.6 \text{ ft. (TDH)}$$

Suction Lift + Other Heads = TDH

Suction Lift + 34 = 37.6

Suction Lift = 3.6 ft.

11. **Answer**   81% efficient

Solve for WHP; solve for pump efficiency.

$$\frac{\text{gpm x hd}}{3960} = \text{whp}$$

$$\frac{355 \text{ x } 41}{3960} = \text{whp}$$

$$3.7 = \text{whp}$$

$$\frac{\text{whp}}{\text{motor eff. x pump eff.}} = \text{mhp}$$

$$\frac{3.7}{.91 \text{ x pump eff.}} = 5$$

$$\text{pump eff.} = .81$$

12. **Answer** $14.59

$$\frac{\text{gpm x hd}}{3960} = \text{whp}$$

$$\frac{700 \times 82}{3960} = \text{whp}$$

$$14.5 = \text{whp}$$

$$\frac{\text{whp}}{\text{w to w eff.}} = \text{mhp}$$

$$\frac{14.5}{.83} = \text{mph}$$

$$17.5 = \text{mph}$$

17.5 hp x .746 Kw/hp x 14 hr. x $.08/Kw.hr. = $14.59

---

13. **Answer**   67% efficiency

Draw Diagram. Convert Kw to mph; solve for TDH, then whp; solve for w-w eff.

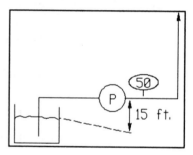

$$5.5 \text{ Kw/30 min.} = 11 \text{ Kw./hr.}$$

$$\frac{11 \text{ Kw/hr}}{.746 \text{ Kw/hp}} = 14.8 \text{ mhp}$$

$$\frac{\text{gpm x hd}}{3960} = \text{whp}$$

$$\frac{300 \times 130.4}{3960} = \text{whp}$$

$$9.9 = \text{whp}$$

$$\frac{\text{wph}}{\text{w-w eff.}} = \text{mhp}$$

$$\frac{9.9}{\text{w-w eff.}} = 14.8$$

$$\text{w-w eff.} = .67$$

14. **Answer**   A. 30 ft.
                  B. 16.7

   A. Drawdown = Static Water Level - Pumping Water Level

       Drawdown = (52 psi x 2.31 ft./psi) - (39 psi x 2.31 ft./psi)

       Drawdown = 30 ft.

   B. Sp.Cap. = gpm/ft. drawdown

       Sp.Cap. = 500/30

       Sp.Cap. = 16.7

---

15. **Answer**   A. 100 ft.
                  B. 1.5

   Note that this data records the Static Water Level as the distance from ground
   surface to the water table, not the depth of static water in the well.

   A. Drawdown = Pumping Water Level - Static Water Level

       Drawdown = 500 - 400

       Drawdown = 100 ft.

   B. Sp.Cap. = gpm/ft. drawdown

       Sp.Cap. = 150/100

       Sp.Cap. = 1.5

---

16. **Answer**   52 psi

   Draw Diagram. Pumping rate and water
   table of 200 ft. below surface not needed for
   this solution. Change pressure to feet of
   water; add to Drawdown; change to psi.

   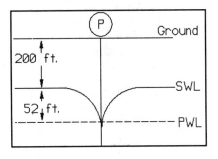

        29.5 psi x 2.31 ft./psi = 68.2 ft.

        68.2 ft. + 52 ft. = 120.2 ft.
                     (water depth-pump off)

        $\dfrac{120.2 \text{ ft.}}{2.31 \text{ ft./psi}}$ = 52 psi

17. **Answer**   A. 73 ft.
                  B. 53 ft.

A. Gage reading (pump off) is equivalent to depth of standing water in the well. Subtract from well depth.

$$55 \text{ psi} \times 2.31 \text{ ft./psi} = 127 \text{ ft. (SWL)}$$

$$200 \text{ ft.} - 127 \text{ ft.} = 73 \text{ ft. to water}$$

B. Gage reading (pump on) is equivalent to pumping water level. Subtract from Static Water Level.

$$32 \text{ psi} \times 2.31 \text{ ft./psi} = 74 \text{ ft.}$$

$$127 \text{ ft.} - 74 \text{ ft.} = 53 \text{ ft. Drawdown}$$

---

18. **Answer**   235 min.

Draw Diagram. Solve for pumping rate; solve for volume of empty half of tank; divide by pumping rate.

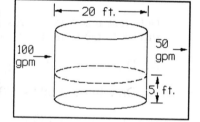

100 gpm in - 50 gpm out = 50 gpm pumping rate

Vol. = .785 d² h

Vol. = .785 x 20² x 5

Vol. = 1570 cu.ft.

Vol. = 11744 gallons

$$\frac{11744 \text{ gal.}}{50 \text{ gal./min.}} = 235 \text{ minutes to fill}$$

---

19. **Answer**   718 gpm

Draw Diagram. Solve for cfs in; solve for cfs drop; rise + drop = gpm pumped out.

Rise = 2 ft./min. 46 seconds; find volume of that 2 ft. depth

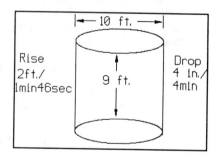

Vol. = .785 d² h

Vol. = .785 x 10² x 2

Vol. = 157 cu.ft.

$$\frac{157 \text{ cu.ft.}}{1.77 \text{ min.}} = 88.7 \text{ cu.ft./min.} = 1.5 \text{ cfs } \underline{\text{in}}$$

Drop = 4 inches/4 min.; find that volume; change to cfs drop.

$$Vol. = .785 \ d^2 \ h$$

$$Vol. = .785 \times 10^2 \times .33$$

$$Vol. = 26 \ cu.ft.$$

$$\frac{26 \ cu.ft.}{4 \ min.} = 6.48 \ cu.ft./min. = .1 \ cfs$$

Rise + Drop = Pumping rate

1.5 + .1 = 1.6 cfs pumping rate

1.6 cfs x 7.48 gal./cu.ft. x 60 sec./min. = 718 gpm

---

20. **Answer**   122.7 gpm

Not all information needed.

$$\frac{3680 \ gal.}{30 \ min.} = 122.7 \ gpm$$

---

21. **Answer**   7.3 psi

Solve for depth of water in tank after pumping; change to pressure.

200 gpm x 60 min./hr. x 4 hr. = 48,000 gal. = 6417 cu.ft.

$$Vol. = .785 \ d^2 \times h$$

$$6417 = .785 \times 22^2 \times h$$

$$17 \ ft. = h$$

$$\frac{17 \ ft. \ depth}{2.31 \ ft./psi} = 7.3 \ psi$$

---

22. **Answer**   2.8 hours

500 gpm <u>in</u> - 200 gpm <u>out</u> = 300 gpm filling rate

empty 2/3 of tank = 50,000 gal.

$$\frac{50,000 \ gal.}{300 \ gpm \ filling} = 166.7 \ minutes \ to \ fill \ tank \ (2.8 \ hours)$$

---

23. **Answer**

Pump will run dry. More water is being pumped out than is coming in.

It is almost impossible to determine a constant regular influent rate for a wastewater lift station. Best solution is to set up pump to respond to water level control devices.

.1 cfs in x 7.48 gal./cu.ft. x 60 sec./min. x 60 min./hr. = 2693 gal./hr. in.

150 gpm out x 20 min/hr. = 3000 gal./hr. out.

---

24. **Answer** 90% of design efficiency

Solve for gpm pumping rate.

Inflow + Drop = Pumping Rate

Influent 110 cfm = 1.83 cfs

Drop 15 inches/4 min. 20 sec. Find volume of drop; change to cfs drop rate.

Vol. = 1 x w x h

Vol. = 11 x 11 x 1.25

Vol. = 151.25 cu.ft. drop - in 4 min. 20 sec.

$$\frac{151.25 \text{ cu.ft.}}{4.33 \text{ min.}} = 24.9 \text{ cu.ft./min.} = .58 \text{ cfs}$$

1.83 cfs in
+ .58 cfs out
1.25 cfs pumping out

1.25 cfs x 7.48 gal./cu.ft. x 60 sec./min. = 1084 gpm
actually being pumped

$$\frac{1084 \text{ gpm pumping rate}}{1200 \text{ gpm design rate}} = .9 = 90\%$$

CHAPTER 7

FRICTION LOSS

This chapter will consider calculations for friction head loss in closed pipe systems. The work is straightforward once the components are understood and inherent weaknesses in theoretical calculation are overcome by adaptation to each particular water system.

Going back a bit, recall that when water flows in a pipeline under pressure, the forward movement of the water is produced by the weight of water behind it, pushing it forward. This pressure, registered in feet, is called Head. It does not matter whether the pressure is originating from a standing tank of water (Pressure Head), or from the slope of the pipe (Elevation Head), or from a combination of the two. In a closed conduit, the water velocity is dependent upon pressure. Since the pipe is completely full and the diameter is constant, the flow is also a result of the pressure driving the water, and is directly proportional to velocity (Q=AV).

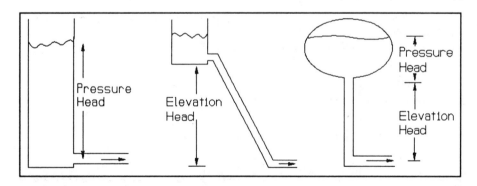

Recall that in a dynamic system, pressure decreases along the length of the pipeline; the Hydraulic Grade Line slopes downward, designating pressure loss from friction created as the water encounters the sides of the pipe. This is the Friction Head Loss which we have been referring to in previous chapters. It is called <u>Major Head Loss</u>, and usually accounts for most of the pressure drop in dynamic water systems. <u>Minor Head Loss</u> is caused by extra turbulence created at bends, fittings, and diameter changes in the pipeline, and in most systems is a lesser component of pressure loss. We will consider Minor Head Loss in Chapter 9.

Since the source of head must provide the needed lift and overcome losses, calculations for head loss are important considerations even before a system is developed, and prior to new construction, pipe replacements, and expansions.

Let us first look at some common sense components of head loss:

ROUGHNESS

The interior surface of a pipe is rough, even when the pipe is new. This interior roughness is dependent upon pipe material, and will increase with age, as the pipe becomes tuberculated and pitted. Normal flow in water pipes is turbulent, and that turbulence increases with pipe roughness; energy is spent, and pressure drops over length.

LENGTH

Friction losses occur with every foot of pipe length, and this must be included in head loss calculations. The longer the pipe, the more head loss.

DIAMETER

Diameter of pipe is also an important factor. Consider the diagram below:

In the large diameter pipe, a certain amount of water is actually touching the sides of the pipe, and is contributing to friction loss. In the small diameter pipe, by comparison, a larger percentage of the water flowing through that pipe is touching the sides. There is less comparable volume on the inside that is not in contact with the pipe wall.

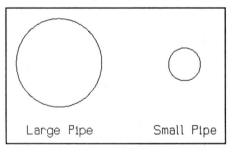

Therefore, small diameter pipes will have more head loss than large diameter pipes.

VELOCITY

Velocity is also a contributing factor. The faster the water flows through the pipe, the more turbulence is created, and the more head loss there will be. So the velocity, or more appropriately, the energy due to velocity. Velocity Head ($V^2/2g$), also has an effect. We can also state it this way: for the same diameter pipe, as the flow increases, head loss increases ($Q = AV$).

In summary, Friction Head Loss (HL$_f$) is dependent upon:

--material and condition of pipe (roughness)
--length of pipe
--diameter of pipe
--velocity head

Good! We have just invented a formula to calculate head loss!

$$HL_f = f \times L \times \frac{1}{d} \times \frac{V^2}{2g}$$

This is the Darcy-Weisbach formula, and has been the basis for head loss calculations since the 19th century. Pipe roughness is designated by a fractional number, f, the friction factor, or Coefficient of Friction. The Darcy-Weisbach formula as such, was meant to apply to the flow of any fluid, and into this friction factor was incorporated not only the degree of roughness, but also an element called the Reynold's Number, which was based on the viscosity of the fluid, and the degree of turbulence of flow. Calculations for f are complex, and the formula for it varies, depending on whether the flow is laminar or turbulent. A special chart, called the Moody Diagram, was developed to help ease calculations, but even this is difficult to interpret, and loses accuracy in systems approaching critical velocity (where laminar flow starts to become turbulent).

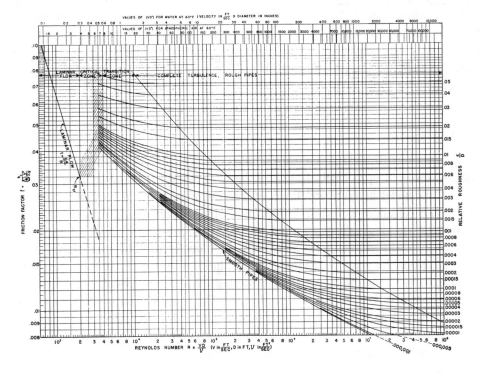

The Moody Diagram

The Darcy-Weisbach formula was flexible enough to adapt to most fluids and velocities, but for popular use in water systems, where the flow is always turbulent, and where the damping effect of viscosity is negligible and unchanging, another formula was developed by Hazen and Williams:

$$Q = .435 \times C \times d^{2.63} \times s^{.54}$$

With a calculator it is easy to compute, and all components can be measured or estimated directly.

With the Hazen-Williams formula came a new designation of pipe roughness. This is the C factor, the modern day Roughness Coefficient. As opposed to Darcy's friction factor (f), C does not vary appreciably with velocity, and by comparing pipe types and ages, it includes only the concept of roughness, ignoring fluid viscosity and Reynold's Number. Equating it numerically to Darcy's friction factor: $f = 200/C^2$ or $C = 14/\sqrt{f}$ This is very approximate, however, for f and C include different fluid properties, and are not truly comparable.

Based on experimentation, accepted tables of C factors have been established for pipe. One is included here (see end of chapter), but there are other similar tables. Estimates of roughness coefficients may be made mentally for pipes not included, and generally C factor decreases by 5-10 with each five years of pipe age. Note that a high C factor means a smooth pipe. A low C factor means a rough pipe. Flow for a newly designed system is often calculated with a C factor of 100, based on averaging it over the life of the pipe system. A knowledge of the particular system is invaluable here, for pipe roughness is very dependent upon water quality, and historical data on the chemical condition of the particular water and its effect on the interior surfaces of the pipes can determine what value is chosen for C.

SLOPE

"Slope" as an English word means the amount of slant, or incline. This applies to water systems also. In open channel systems, such as aqueducts or sewer pipe, where the water is not under pressure and flows by gravity, slope is the amount of incline of the pipe, and is calculated as feet of drop per foot of pipe length (ft./ft.). In sewer systems, the slope of the pipe is designed to be just enough to overcome frictional losses, so that the velocity (and flow) will remain constant; the water keeps flowing, and solids will not settle out along the way. The same principle holds true in a closed pipe system under pressure. With these pipes we are considering a horizontal conduit which encounters "feet" of pressure loss for every foot of pipe. The pressure loss would slow down the velocity, restricting flow, and so must be overcome. It is the same as the slope in an open system, but instead of actually slanting the pipe to overcome the losses, we add a source of pressure to overcome friction. It is still called "slope": feet of head loss per foot of pipe (ft./ft.), which must be overcome. To calculate for it, we divide pressure loss by length of pipe:

$$s = HL_f /L$$

Now that we have identified all the components of the Darcy-Weisbach and the Hazen-Williams formulas, let's look at them again. They seem quite different from each other. Actually, they are not.

Derivation:

Take Darcy's:

$$HL_f = f \ x \ L \ x \ \frac{1}{d} \ x \ \frac{V^2}{2g}$$

We will rearrange it; move L to the left.

$$\frac{HL_f}{L} = f \ x \ \frac{1}{d} \ x \ \frac{V^2}{2g}$$

$\frac{HL}{L}$ is slope; move it back to the right side as slope; move V to the left side; invert the whole formula:

$$V^2 = s \ x \ \frac{d}{f} \ x \ 2g$$

Take the square root of every value:

$$V = \frac{s^{.5}}{f^{.5}} \ x \ d^{.5} \ x \ 2g^{.5}$$

$$V = \frac{Q}{A}$$

Equate V to Q:

$$Q = A \ x \ \frac{s^{.5}}{f^{.5}} \ x \ d^{.5} \ x \ 2g^{.5}$$

Enter numerical values for A & 2g:

$$Q = .785 \ d^2 \ x \ \frac{s^{.5}}{f^{.5}} \ x \ d^{.5} \ x \ 64.4^{.5}$$

Change f to C; condense: $(f = 200/C^2)$

$$Q = .445 \ x \ C \ x \ d^{2.5} \ x \ s^{.5}$$

Original version given for Hazen-Williams formula is very similar:

$$Q = .435 \ x \ C \ x \ d^{2.63} \ x \ s^{.54}$$

With this rough conversion of one formula to another, and understanding that f and C factors are not accurately equatable, the point is that a quite formidable looking formula such as the Hazen-Williams, also originates from the concepts of roughness, length, diameter and velocity head, which are the basic components of head loss.

There are many variations of the Hazen-Williams formula; some have slightly different exponents; some equate to velocity instead of to flow; some are specialized for open channel flow; other hydraulics experts have developed their own variations, and use them under their own names, but it is easy to recognize the general format of the equation, and they all calculate out to yield the same approximate value of any one dimension. It is important to stress "approximate", for a few inherent weaknesses occur in any of these formulas.

There is no problem with flow, velocity, slope and length. If these are to be the known factors, they can be measured, metered, read from gages. The weakest point in the formula is the Roughness Coefficient, C. If this is to be a known factor for calculation, at best it can only be estimated, based on pipe type, age, and knowledge of water quality. The only way to accurately determine the amount of roughness in a pipe is to measure all the other components of the formula, and then calculate for C. This is done by utilities to determine when pipe needs to be cleaned.

For example: Isolate a pipe section. Attach gages to two hydrants, one downstream from the other. Open a third hydrant downstream from both of these to induce flow. Use a pitot gage to measure flow at the open hydrant. Use Hazen-Williams formula to calculate for C factor.

Pipe diameter can also lead to inaccurate calculations. Inside diameter is actually what should be used (that's where the water is), but even using rated, or nominal diameter, little accuracy is lost. Most important is knowledge of the particular system's water quality. How much rust buildup, chemical post-precipitation, or scale formation have occurred since that pipe was laid? A six inch diameter pipe can quickly become a two inch diameter pipe if the water quality is poor. In how many systems can we assume that the diameter of the pipe is the same as when it was new? This is where the operator's knowledge of his treatment system counts.

In addition, we are always assuming steady state flow; for the particular section we are observing for calculation, the water that enters the system, leaves the system. We must hold this true in order to keep calculations reasonably simple. But how many systems have no leaks?

For field use, as an alternate to calculating the Hazen-Williams formula, an alignment chart has become popular, and can be used with reasonable accuracy. This chart is of a special type called a nomograph. It may yield an answer with one or more of its dimensions unknown (see next page). Inspecting it, we note dimensions of discharge, pipe diameter, water velocity, C factor, and head loss in ft./1000 ft. There is also a Pivot Line, for getting from one side to the other.

DIRECTIONS FOR USING
HAZEN-WILLIAMS ALIGNMENT CHART

For determining head loss, find the proper discharge and line it up with the pipe diameter. Extend the line across to the Pivot Line, and mark the point where the Pivot Line was crossed. Line up that point with the proper C factor on the other side, and extend it to the head loss.

Velocity is determined from a line which passes through flow and diameter. Pivot Line is not needed for this. Pivot Line is needed only for finding C factor or head loss.

Values of C for Hazen-Williams Formula

Type of Pipe	C
Asbestos Cement	140
Brass	140
Brick Sewer	100
Cast Iron	
New, unlined	130
New, lined	140
10 years old	110
20 years old	90
Ductile Iron, (cement lined)	140
Concrete or Concrete Lined	
Smooth, steel forms	140
Wooden forms	120
Rough	110
Copper	140
Fire Hose (rubber lined)	135
Galvanized Iron	120
Glass	140
Lead	130
Masonry Conduit	130
Plastic	150
Steel	
Coal-tar enamel lined	150
New Unlined	140
Riveted	110
Tin	130
Vitrified	120
Wood Stave	120

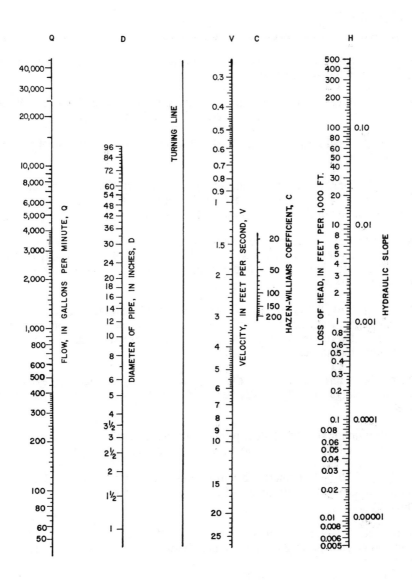

Hazen-Williams Alignment Chart

PROBLEMS

1. A 36 inch diameter pipe carries a flow of 800 gpm. If this pipe has a C value of 100, what is the slope (ft./ft.)?

2. A flow of 10 cfs occurs in a rough concrete pipe of 30 inch diameter. What is the total head loss due to friction in a 2000 ft. length of the pipe?

3. What is the head loss in 350 ft. of 10 inch diameter new welded steel pipe carrying .8 MGD?

4. A 4 inch diameter cast iron main 40 years old and 1000 ft. long carries .2 cfs through a section of town. How much pressure (ft.) must be provided by the utility to maintain the flow for this section alone?

5. An elevated tank 50 ft. off the ground supplies your house with water. The water flows at 2 cfs through a 10 inch main 500 ft. long (10 year old CIP). How deep does the water have to be in the tank in order to provide your house with a pressure of 40 psi?

6. A 9000 ft. long, 18 inch diameter AC pipeline 10 years old delivers 4.5 MGD to a community which is 100 ft. higher in elevation than the treatment plant. The pressure delivered must be at least 60 psi. What pressure must be provided at the treatment plant to achieve this (psi)?

7. Water flows through a 6 inch diameter horizontal galvanized pipe 1000 ft. long with an average velocity of 8 ft./sec. What is the pressure drop (psi) across this line?

8. What minimum size lined steel pipe should be used to carry a flow of at least 10 cfs. if the total friction loss is to be 25 ft. over each mile of pipe length?

9. In a test of 20 inch diameter cast iron pipe, the flow is steady at 2.2 cfs, and the Hydraulic Grade Line dropped 4 ft. in a length of 8000 ft. of pipe. What is the value of C?

10. A 10 inch diameter horizontal concrete pipe (C=120) carries water at 1500 gpm from point 1 to point 2. The distance between the two points is 1000 ft. The pressure at point 1 is 80 psi.
    A. What is the pressure at point 2?
    B. What is the difference in the water velocity between the two points?

11. Water flows through a 12 inch diameter galvanized iron pipe 800 ft. long with a velocity of 9 ft./sec. The pipe outlet is 50 ft. above the pipe inlet. What is the pressure difference between inlet and outlet (psi)?

12. What lost head per foot of pipe in a new 16 inch diameter cast iron pipe will create the same flow as occurs in a new 20 inch diameter (CIP) pipe with a drop in the HGL of one foot per thousand feet?

13. Determine the discharge through a new cement lined steel pipe of 1 ft. diameter if the pressure drop across the length of pipe is 2 psi per 1000 ft.?

14. When first installed between two reservoirs a 4 inch diameter cast iron pipe 6000 ft. long conveyed 1 cfs of water. If after 15 years, a chemical deposit had reduced the effective diameter to 3 inches, what then would be the flow rate? (assume no change in C value)

15. What percent error is introduced into Q when C is misjudged by 20%?

16. A. What size pipe would you require to transport 3900 gph of water if the head loss in 570 ft. of the masonry pipe had to be less than or equal to 1.18 ft.?
    B. If you had to maintain at least a velocity of 2.5 ft./sec., how much head loss would be in the pipe?

17. Given the 800 ft. pipe below, calculate the friction head loss A-B(ft.), (C=100)

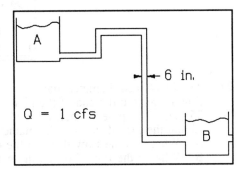

18. With 1 cfs flowing, what is the required total head to maintain flow at steady state in the system shown (assume 20 year old CIP)?

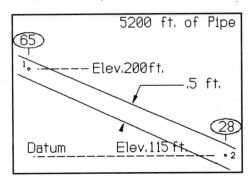

19. A sewer carries a full flow of 5.2 cfs. If C=100 and slope = .003 ft./ft., what is the water velocity?

20. A pressure gage at point 1 on a horizontal 8 inch diameter pipeline reads 45 psi. At point 2 on the same pipe, 100 ft. downstream, another pressure gage shows the same reading. Assuming that both gages are functioning correctly, what is the flow?

21. You have 5000 ft. of 12 inch diameter pipe with a C factor of 110, carrying a flow of 1.7 MGD.
    A. What is the total head loss?
    B. If you replaced it with a 16 inch diameter pipe, with the same C factor, at the same flow, what would be the pressure drop across this line (ft.)?

22. A horizontal high pressure line feeds water across an osmotic membrane at 400 psi. The head loss across the membrane is 700 ft. What is the pressure on the downstream side of the membrane (psi)?

23. Given the diagram to the right, find:
    A. Tank diameter.
    B. Condition of pipe (C factor).
    C. Gage pressure.

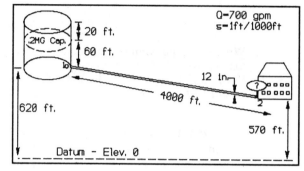

24. An old 14 inch diameter main (C=60) carries water 2000 ft. to a pond where the flow is free out the end of the pipe. A gage on the upstream end of the pipe reads 35 psi. The pipe is then cleaned and has a new smooth plastic liner inserted. Lining the pipe decreases the diameter to 13 inches.
    A. What was the flow through the old pipe?
    B. What is the flow through the new pipe?

SOLUTIONS

1. **Answer**   .00001 ft./ft.

   Change gpm to cfs; solve for slope.

   $$\frac{800 \text{ gpm}}{60 \text{ sec./min x } 7.48 \text{ gal./cu.ft.}} = 1.78 \text{ cfs}$$

   $$Q = .435 \ C \ d^{2.63} \ s^{.54}$$

   $$1.78 = .435 \ 100 \ 3^{2.63} \ s^{.54}$$

   $$.00001 = s$$

2. **Answer**   1.2 ft.

   Rough concrete C=110; 30 inches = 2.5 ft.

   $$Q = .435 \ C \ d^{2.63} \ s^{.54}$$

   $$10 = .435 \ 110 \ 2.5^{2.63} \ s^{.54}$$

   $$.0006 = s$$

$$s = \frac{HL}{L}$$

$$.0006 = \frac{HL}{2000}$$

$$1.2 \text{ ft.} = HL$$

---

3. **Answer**   .64 ft.

Change MGD to cfs; 10 in. = .83 ft.; Welded steel C=140.

$$.8 \text{ MGD} \times 1.55 = 1.24 \text{ cfs}$$

$$Q = .435 \ C \ d^{2.63} \ s^{.54}$$

$$1.24 = .435 \ 140 \ .83^{2.63} \ s^{.54}$$

$$.002 = s$$

$$HL = .002 \text{ ft./ft.} \times 350 \text{ ft.} = .64 \text{ ft.}$$

---

4. **Answer**   37.5 ft.

CIP 40 yr. old C=50

$$Q = .435 \ C \ d^{2.63} \ s^{.54}$$

$$.2 = .435 \ 50 \ .33^{2.63} \ s^{.54}$$

$$.0375 = s$$

$$HL = .0375 \text{ ft./ft.} \times 1000 \text{ ft.} = 37.5 \text{ ft.}$$

---

5. **Answer**   45.9 ft.

Draw Diagram. Pressure at house (40 psi) is equivalent to 92.4 ft. With Bernoulli's formula, solve for depth needed; solve for HL; add to depth.

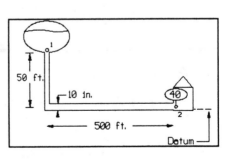

$$Ph_1 + Z_1 = Ph_2 + Z_2 + HL$$

$$Ph + 50 = 92.4 + 0 + HL$$

$$Ph = 42.4 + HL$$

$$Q = .435 \ C \ d^{2.63} \ s^{.54}$$

$$2 = .435 \ 110 \ .83^{2.63} \ s^{.54}$$

$$.007 = s$$

$$HL = .007 \text{ ft./ft.} \times 500 \text{ ft.} = 3.5 \text{ ft.}$$

$$\begin{array}{r} 42.4 \text{ ft.} \\ + 3.5 \text{ ft.} \\ \hline 45.9 \text{ ft.} \end{array}$$

---

6. **Answer**   114.6 psi

Draw Diagram.   10 yr. old AC pipe C=130. Treatment plant must deliver 60 psi plus 100 ft. elevation, plus make up HL; solve for HL; add:

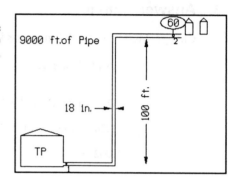

$$4.5 \text{ MGD} \times 1.55 \text{ cfs/MGD} = 6.98 \text{ cfs}$$

$$60 \text{ psi} \times 2.31 \text{ ft./psi} = 138.6 \text{ ft.}$$

$$Ph_1 + Z_1 = Ph_2 + Z_2 + HL$$

$$Ph_1 + 0 = 138.6 + 100 + HL$$

$$Ph_1 = 238.6 + HL$$

$$Q = .435 \ C \ d^{2.63} \ s^{.54}$$

$$6.98 = .435 \ \ 130 \ \ 1.5^{2.63} \ \ s^{.54}$$

$$.003 = s$$

$$HL = .003 \text{ ft./ft.} \times 9000 \text{ ft.} = 26 \text{ ft.}$$

$$238.6 \text{ ft.} + 26 \text{ ft.} = 264.6 \text{ ft.}$$

$$\frac{264.6 \text{ ft.}}{2.31 \text{ ft./psi}} = 114.6 \text{ psi}$$

---

7. **Answer**   19.5 psi

Galv. pipe C=120. Solve for flow; solve for HL.

$$Q = A \ V$$

$$Q = .785 \times .5^2 \times 8$$

$$Q = 1.57 \text{ cfs}$$

$$Q = .435 \ C \ d^{2.63} \ s^{.54}$$

$$1.57 = .435 \ 120 \ .5^{2.63} \ s^{.54}$$

$$.045 = s$$

HL = .045 ft./ft. x 1000 ft. = 45 ft.

$$\frac{45 \ ft.}{2.31 \ ft./psi} \ 19.5 \ psi$$

---

8. **Answer**   18 inch diameter pipe

slope 25 ft./5280 ft. = .0047

lined steel pipe C=150

$$Q = .435 \ C \ d^{2.63} \ s^{.54}$$

$$10 = .435 \ 150 \ d^{2.63} \ .0047^{.54}$$

1.47 ft. = d

use 18 inch pipe

---

9. **Answer**   80

slope = 4 ft./8000 ft. = .0005

20 inches = 1.67 ft.

$$Q = .435 \ C \ d^{2.63} \ s^{.54}$$

$$2.2 = .435 \ C \ 1.67^{2.63} \ .0005^{.54}$$

80 = C

---

10. **Answer**   A. 73.5 psi
    B. Velocity is the same at both points.
        (Flow is steady - area does not change; Q=AV.)

Draw Diagram.   Change psi to ft.;
change gpm to cfs; solve for HL;
subtract.

80 psi x 2.31 ft./psi = 184.8 ft.

$$\frac{1500 \ gpm}{60 \ sec./min. \ x \ 7.48 \ gal./cuft.} = 3.3 \ cfs$$

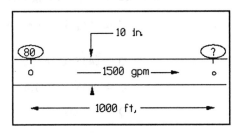

$$Q = .435 \ C \ d^{2.63} \ s^{.54}$$

$$3.3 = .435 \ 120 \ .83^{2.63} \ s^{.54}$$

$$.015 = s$$

$$HL = .015 \ \text{ft./ft.} \times 1000 \ \text{ft.} = 15 \ \text{ft.}$$

$$184.8 \ \text{ft. (point 1)} - 15 \ \text{ft. (HL)} = 169.8 \ \text{ft.} = 73.5 \ \text{psi}$$

---

11. **Answer**   30.3 psi difference

Draw Diagram. As pipe slopes up, pressure drops-equivalent to 50 ft.; pressure also drops from HL; solve for Q; solve for HL; add to elevation.

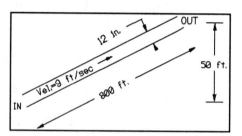

Galvanized C=120

$$Q = A \ V$$

$$Q = .785 \times 1^2 \times 9$$

$$Q = 7.1 \ \text{cfs}$$

$$Q = .435 \ C \ d^{2.63} \ s^{.54}$$

$$7.1 = .435 \ 120 \ 1 \ s^{.54}$$

$$.025 = s$$

$$HL = .025 \ \text{ft./ft.} \times 800 \ \text{ft.} = 20 \ \text{ft.}$$

$$50 \ \text{ft. (elev.)} + 20 \ \text{ft. (HL)} = 70 \ \text{ft. pressure difference}$$

$$70 \ \text{ft.} = 30.3 \ \text{psi}$$

---

12. **Answer**   .003 ft./ft.

CIP C=130
Find flow in 20 in. pipe; using that flow, solve for s in 16 in. pipe

$$HGL \ \text{drop} \ 1 \ \text{ft./1000 ft.} = .001 = s$$

$$Q = .435 \ C \ d^{2.63} \ s^{.54}$$

$$Q = .435 \ 130 \ 1.67^{2.63} \ .001^{.54}$$

$$Q = 5.2 \ \text{cfs}$$

$$Q = .435 \ C \ d^{2.63} \ s^{.54}$$

$$5.2 = .435 \ 130 \ 1.33^{2.63} \ s^{.54}$$

$$.003 = s$$

Note: Since the flow in both pipes is the same, <u>the Total HL will be the same in both</u>. The smaller pipe has a greater HL per foot of pipe - and must then be a shorter pipe.

---

13. **Answer**   3.33 cfs

New cement steel C=140

HL = 2 psi/1000 ft. = 4.62 ft./1000 ft. = .0046 ft./ft.

$$Q = .435 \ C \ d^{2.63} \ s^{.54}$$

$$Q = .435 \ 140 \ 1 \ .0046^{.54}$$

$$Q = 3.33 \ cfs$$

---

14. **Answer**   .48 cfs

CIP C=140; solve for HL in the new pipe; with narrowed pipe, at the same HL, solve for flow. This is not a steady state system. The flow will change-because the pipe narrowed.

New Pipe

$$Q = .435 \ C \ d^{2.63} \ s^{.54}$$

$$1 = .435 \ 140 \ .33^{2.63} \ s^{.54}$$

$$.1097 = s$$

Narrowed Pipe

$$Q = .435 \ C \ d^{2.63} \ s^{.54}$$

$$Q = .435 \ 140 \ .25^{2.63} \ .1097^{.54}$$

$$Q = .48 \ cfs$$

---

15. **Answer**   20%

Test it out with 2 pipes; keep all quantities the same - change C by 20%; solve for flow.

---

16. **Answer**   A. 6 inch diam.
                  B. ?

Masonry conduit C=130; change gph to cfs.

   3900 gph/7.48 x 60 x 60 = .145 cfs

A. Solve for diam. First find s.   1.18/570 = .0021 ft./ft.

   $Q = .435 \ C \ d^{2.63} \ s^{.54}$

   $.145 = .435 \ 130 \ d^{2.63} \ .0021^{.54}$

   .367 ft. = d

   4.4 in. = d

   6 in. pipe would be chosen

B. Solve for velocity with 6 inch pipe at this flow.

   $Q = A \ V$

   $.145 = .785 \ x \ .5^2 \ V$

   .739 ft./s = V

A decision must be made.
   a. Decrease pipe size to increase velocity - but Q will go down-is that acceptable in this system?
   b. Maintain pipe size and increase flow - velocity will go up, but so will HL. - is that acceptable?
   c. Get a smoother pipe
A compromise must be made

---

17. **Answer**   21.6 ft.

Without head loss in the pipe, the water level in B would be the same as A. Imagine Res. B as a large, wide piezometer; its water level will register the pressure there, in feet. Therefore the difference in the water levels in the two reservoirs - is the HL.

   $Q = .435 \ C \ d^{2.63} \ s^{.54}$

   $1 = .435 \ 100 \ .5^{2.63} \ s^{.54}$

   .027 = s

   HL = .027 ft./ft. x 800 ft. = 21.6 ft. difference in elevation of the water levels.

18. **Answer**   170 ft.

Change psi to feet:

> Point 1:   65 psi = 150.15 ft.
>
> Point 2:   28 psi = 64.68 ft.
>
> Elevation difference: 200 ft. - 115 ft. = 85 ft.

Set up Bernoulli's formula: Velocity Heads cancel.

> $Ph_1 + Z_1 = Ph_2 + Z_2 + HL$
>
> $150.15 + 200 = 64.68 + 115 + HL$
>
> 170.5 ft. $= HL$

The problem could also have been solved using Hazen-Williams formula.

> $Q = .435 \ C \ d^{2.63} \ s^{.54}$
>
> $1 = .435 \ 90 \ .5^{2.63} \ s^{.54}$
>
> $.0329 = s$
>
> $HL = .0329$ ft./ft. x 5200 ft. $= 170$ ft.

---

19. **Answer**   3 ft./sec.

Solve for diameter; solve for velocity.

> $Q = .435 \ C \ d^{2.63} \ s^{.54}$
>
> $5.28 = .435 \ 100 \ d^{2.63} \ .003^{.54}$
>
> 1.48 ft. $= d$
>
> 18 in. $= d$
>
> $Q = A \ V$
>
> $5.28 = .785$ x $1.5^2 \ V$
>
> 3 ft./sec. $= V$

---

20. **Answer**   The gages register no head loss - therefore, no flow.

---

21. **Answer**   A. 23.4 ft.
                        B. 5.8 ft.

         1.7 MGD x 1.55 cfs/MGD = 2.64 cfs

A.
$$Q = .435 \; C \; d^{2.63} \; s^{.54}$$

$$2.64 = .435 \;\; 110 \;\; 1 \;\; s^{.54}$$

$$.0047 = s$$

$$HL = .0047 \text{ ft./ft.} \times 5000 \text{ ft.} = 23.4 \text{ ft.}$$

B. Insert the 16 inch line; solve for HL.

$$Q = .435 \; C \; d^{2.63} \; s^{.54}$$

$$2.64 = .435 \;\; 110 \;\; 1.33^{2.63} \;\; s^{.54}$$

$$.0012 = s$$

$$HL = .0012 \text{ ft./ft.} \times 5000 \text{ ft.} = 5.8 \text{ ft. HL}$$

---

22. **Answer**   97 psi

Change pressure to feet; subtract ft. of HL; change back to psi.

$$400 \text{ psi} \times 2.31 \text{ ft./psi} = 924 \text{ ft.}$$

$$924 \text{ ft.} - 700 \text{ ft. HL} = 224 \text{ ft. remaining}$$

$$\frac{224 \text{ ft.}}{2.31 \text{ ft./psi}} = 97 \text{ psi}$$

---

23. **Answer**   A. 20.6 ft.
                        B. 150
                        C. 45.9 psi

A.
$$\text{Volume} = .785 \times d^2 \times h \times 7.48 \text{ gal./cu.ft.}$$

$$200,000 = .785 \times d^2 \times 80 \times 7.48$$

$$20.6 \text{ ft.} = d$$

B. Change gpm to cfs; solve for C.

$$\frac{700 \text{ gpm}}{60 \times 7.48} = 1.56 \text{ cfs}$$

$$Q = .435 \; C \; d^{2.63} \; s^{.54}$$

$$1.56 = .435 \;\; C \;\; 1 \;\; .001^{.54}$$

$$150 = C$$

C. Set up Bernoulli's formula;
   HL = .5 ft./1000 for a length of 4000 ft. = 2 ft. HL; Velocity Heads cancel.

$$Ph_1 + Z_1 = Ph_2 + Z_2 + HL$$

$$60 + 620 = Ph_2 + 570 + 4$$

$$106 \text{ ft.} = Ph_2$$

$$45.9 \text{ psi} = \text{pressure}$$

---

24. **Answer**   A. 6.9 cfs
                    B. 14.2 cfs

35 psi = 80.85 ft. Since head is 0 at downstream end, this is the HL.
80.85/2000 = .04 = s
Plastic C=150
14 in. = 1.167 ft.
13 in. = 1.083 ft.

A. Old Pipe: $Q = .435 \; C \; d^{2.63} \; s^{.54}$

$$Q = .435 \quad 60 \quad 1.167^{2.63} \quad .04^{.54}$$

$$Q = 6.9 \text{ cfs}$$

B. New Pipe: $Q = .435 \; C \; d^{2.63} \; s^{.54}$

$$Q = .435 \quad 150 \quad 1.083^{2.63} \quad .04^{.54}$$

$$Q = 14.2 \text{ cfs}$$

CHAPTER 8

COMPOUND PIPES

Municipal water distribution systems are not built with separate single pipes extending from treatment plant to customer's residence. Rather, a network of pipes is laid under the streets - different diameters, lengths, ages and materials. Some branch and loop; some are connected end to end. These more complex arrangements are referred to as <u>Compound Pipe Systems</u>. There are two basic types with which we will consider some calculations:

<u>Pipes in Series</u> - two or more different pipes, laid end to end.
<u>Pipes in Parallel</u> - two or more pipes laid side by side, with the flow splitting among them.

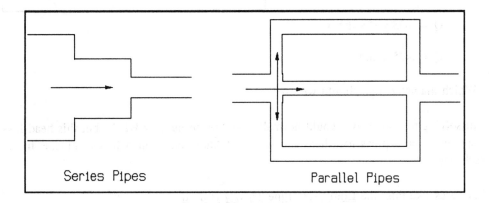

Series Pipes   Parallel Pipes

CALCULATING WITH PIPES IN SERIES
Basic Principles
When pipes are laid in series:

A) The same flow passes through all of them.

B) The total head loss is the sum of the head losses of all of the component pipes.

Situation I

In series systems where the flow is given, and the total head loss is the unknown:

1) Use the Hazen-Williams formula to solve for the slope and the head loss of each, as if they were separate pipes.

2) Add up the head losses to get the total head loss.

Situation II

In systems where the total head loss is given, and the flow is the unknown, it is a little more complicated. Consider the system below:

Total HL$_f$ = 50 ft.

C = 100 all pipes

Find Q

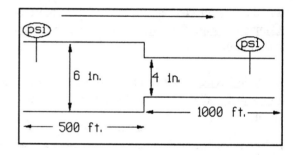

We must solve for the flow.
Using the Hazen-Williams formula:

$$Q = .435 \times C \times d^{2.63} \times s^{.54}$$

$$Q = .435 \times 100 \times ? \times ?$$

--Which diameter pipe should we use?

--In solving for slope, we would normally use the formula s=HL/L, but this head loss applies to both pipes together, and we don't know how much loss originates from each one.

For cases like this, the <u>Equivalent Pipe Theory is used</u>.

We will mentally create one single <u>Equivalent Pipe</u> which will carry the correct flow, because the head loss through it is the same as that in our actual system. We can create our equivalent pipe with any diameter and C factor we wish, just as long as we use those same dimensions for it all the way through to the end. From the diagram, if we make our equivalent pipe long enough to yield 50 ft. of head loss, then the flow passing through this pipe will be the same as the flow in our actual system.

<u>Whatever the flow and total head loss may be, they are the same in the actual system, and in the equivalent pipe.</u>

Now, this equivalent pipe must have the right length, so that it will allow the correct flow through, which yields the correct head loss (our given head loss).

1) Start by looking back at our actual system diagrammed. We must first have a flow and a head loss for this system, which we can apply to the equivalent pipe, in order to solve for a length. <u>Assume</u> a flow -any flow.

2) Using Hazen-Williams formula, solve for slope, and head loss (based on the assumed flow) for each pipe component of it.

3) Add them up to get the total head loss through the system. This is not the true head loss (true head loss is 50 ft.), but it is correct for our assumed flow.

4) Recalling that flow and head loss are the same in the equivalent pipe as in the actual system - even if that flow is an assumed one - construct the equivalent pipe. Chose any diameter and C factor; use the assumed flow and solve for the slope.

5) Now find the length of the equivalent pipe. Using s=HL/L, insert the head loss just acquired from calculation with the assumed flow. Insert slope just acquired from calculation with the assumed flow. Solve for length.

6) Since both head loss and slope were derived from calculations based on the same assumed flow, they have a specific numerical relation to each other. This relation would be true, and would be the same, no matter what we had chosen to assume as the flow. This relation - is the length of our equivalent pipe. It would be the same length no matter what flow we had used - even if it had been the true flow of our actual system. This length, therefore, is a <u>real</u> length!

7) Use it! Again, with s=HL/L, insert the length of the equivalent pipe. Go back to the actual system, and take the original correct head loss (given 50 ft.). Insert that. Solve for slope.

8) Using Hazen-Williams formula, solve for the flow actually passing through that pipe, which is the flow of our original system.

Check your answer. Insert this calculated flow into Hazen-Williams formula for the actual system; solve for slope and head loss of each pipe component; add them up. It should come to 50 ft. total head loss.

Equivalent pipes may be substituted for actual pipes in any system.

CALCULATING WITH PIPES IN PARALLEL
Basic Principles
When pipes are laid in parallel:

A) The <u>flow splits</u> among them. Least flow is admitted to the pipe with the most resistance (smaller diameter, longer, rougher). The greatest flow will pass through the pipe with the least resistance (wider, shorter, smoother).

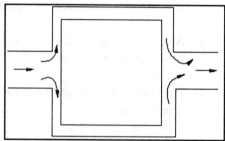

B) The <u>head loss</u> in each pipe component of a parallel system - <u>is the same</u>. The flow split makes it so. Taking the path of least resistance, just enough flow is admitted to each pipe, so that the head losses are balanced. Each pipe has the same head loss.

C) The head loss through the entire parallel system is the same as the head loss through any one of its pipe components. If one equivalent pipe were substituted for this entire system, and the approaching flow were passed through it, the head loss registered would be identical to that registered through each component pipe of the parallel system. That is what makes this pipe equivalent.

D) In parallel pipe systems, the <u>head losses are not added up</u>. The flows are added up.

Situation I
In parallel pipe systems, when the actual head loss is given (across one pipe segment, or across the whole system), and the approaching flow, as well as the component flows, are unknown:

1) Take one pipe of the set, divide head loss by length to get slope.

2) Use Hazen-Williams formula, solve for flow in that pipe.

3) Do the same with the other component pipes.

4) Add up the flows to get the approaching flow.

Situation II

In parallel systems where the approaching flow is given, but the flows and head losses of the components are to be found:

1) We can replace this entire system with one equivalent pipe.

2) Assume a head loss; using Hazen-Williams formula and our actual system, solve for flow in each pipe based on the assumed head loss.

3) Add up these flows to get total flow.

4) Create the equivalent pipe. Choose any diameter and C factor. Calculate what length pipe will give us this total flow for this assumed head loss. To do this use Hazen-Williams formula and the new total flow to solve for slope - in our equivalent pipe.

5) Now we have a slope for our equivalent pipe which is based on our assumed head loss. Using both of these, solve for equivalent pipe length (s=HL/L). This is the real length.

6) We can now replace our parallel system with this one equivalent pipe. It has all the needed dimensions.

7) Take the actual flow approaching the system. Solve for slope, and head loss, in the equivalent pipe. This head loss is the same as the head loss in each segment of the original parallel system.

8) To solve for flows in the separate pipe components of the parallel system, use this head loss to get slope, and then flow, for each.

To check your answer, add up the flows. They should sum up to the approaching flow, which was originally given.

There is a short cut which may be used in a case like this. After proceeding through step 3, note that we have created flows for our actual system which were based on an assumed head loss, and so are not numerically true. However, the percentage flow split among the pipes is valid. We can take each component flow derived from the assumed head loss, and calculate its percentage of the total flow. Multiply each percentage by the actual approaching flow to get the true component flows in the parallel system. To solve for head loss in each pipe, use Hazen-Williams formula to derive slope, and then head loss.

The type of calculations covered in this chapter are the basis for solving piping network problems. An analysis of flows and pipe conditions in an entire section of a city water distribution system may be required for planning system expansions, maintenance, major replacements, and for troubleshooting low pressure problems.

There are several solution methods in use. The equivalent pipe method which we have chosen is one of the simplest. Another method of network analysis is the Hardy-Cross system. For instance, for a parallel pipe system, assign arbitrary positive values to flows in the clockwise direction, and negative values to flows in the counterclockwise direction, solving for head losses, and then adding them algebraically (one is positive, one is negative). When this yields a result of zero, head losses are equal, and the correct flow has been chosen. However, this method involves many trial runs to determine the actual flows and is less direct than the equivalent pipe method.

Manual solution of real network problems can be a large undertaking. The use of the Hazen-Williams alignment nomograph can speed up the calculations, and for systems involving more than a few pipes, computer systems which automate the process are in widespread use.

Steps for solution have been outlined in the chapter. Rather than attempting more detailed explanations here, try working through some of the following problems.

PROBLEMS

1. Water is flowing at 1 cfs from tank A to tank B. Tank A is 20 ft. deep. How deep is tank B?

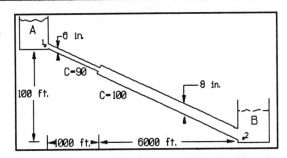

2. One thousand feet of 8 inch diameter pipe is extended by 1200 ft. of 10 inch diameter pipe. The velocity through the 10 inch pipe is 2 ft./sec. If C=100 for both pipes, what is the head loss in this system (ft.)?

3. A cast iron pipeline, C=100, connects points A,B,C,D, which have respective elevations of 347 ft., 357 ft., 361 ft., and 351 ft. Pipe diameters and lengths are:

   A to B   6 inches, 450 ft.
   B to C   8 inches, 320 ft.
   C to D   10 inches, 280 ft.

   Flow is .5 cfs and pressure at A is 73 psi. Determine the pressure at Point E, located on pipe BC, and 100 ft. downstream from B.

4. A 10 inch diameter pipe 1400 ft. long is connected to an 8 inch diameter pipe 1200 ft. long, and then to a 6 inch diameter pipe, 1500 ft. long, all in series (C=100 all pipes).
   A. Determine the length of an equivalent 14 inch pipe.
   B. Determine the length of an equivalent 6 inch pipe.

5. A 12 inch diameter cast iron pipe ten years old carries a flow of 1 MGD for 6000 ft. After a main break, half of its length was replaced by new 12 inch ductile iron pipe. What is the head loss through the entire length of pipe at this flow (ft.)?

6. Given the system to the right:
   A. With the valve closed, what is the head loss in pipe B?
   B. With the valve open, what is the head loss in B?
   C. With the valve open, what is the head loss across the entire system?

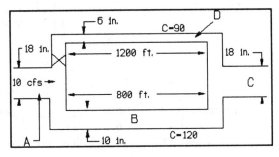

7. Two pipes arranged in parallel admit a total flow of 3 cfs. One pipe has a 12 inch diameter and is 1500 ft. long. The other has a 6 inch diameter and is 2000 ft. long. What is the head loss across this system (C=100 all pipes)?

8. Find the water velocity in the 6 inch diameter pipe.

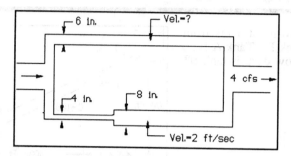

9. Three thousand feet of 6 inch diameter pipe is connected in series to 4000 ft. of pipe whose diameter is not known. If the flow through the system is .3 cfs and the pressure drop across the system is 20 psi, what is the unknown diameter (C=80 all pipes)?

10. A reservoir at 700 ft. is to supply a second reservoir at 460 ft. The reservoirs are connected by 1300 ft. of 24 inch diameter new cast iron pipe and 2000 ft. of 20 inch diameter new cast iron pipe, in series. What will be the discharge delivered from the upper reservoir to the lower one?

11. Which of these systems has the greater capacity (C=120 all pipes)?

    A. <u>9000 ft. (16 inches)</u>   6000 ft. (12 inches)   3000 ft. (10 inches)

    B. <u>11,000 ft. (18 inches)</u>   5000 ft. (8 inches)   2500 ft. (10 inches)

                               7000 ft. (10 inches)

12. Given the system to the right, what is the difference in elevation of the water surfaces in the two piezometers?

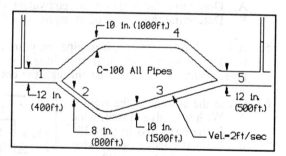

13. Three pipes are arranged in parallel:

    Top section: 4 inch diameter, C=140, 570 ft. long
    Mid section: 6 inch diameter, C=80, 900 ft. long.
    Bottom section: 8 inch diameter, C=100, 1200 ft. long

    The approaching flow is 3000 gpm.
    A. Compute the equivalent length of a 12 inch diameter pipe with a C=100
    B. What is the head loss through this system?

14. Compound pipe ABCD consists of 10,000 ft. of 20 inch, 8000 ft. of 16 inch, and L ft. of 12 inch. all in series. (C=120 all three).
    A. What length of L will make the ABCD pipes equivalent to a 15 inch pipe, 16500 ft. long (C=100).
    B. If the length of the 12 inch pipe were 3000 ft., what flow would occur for a lost head of 135 ft. from A to D?

15. Calculate the flow through the system below:

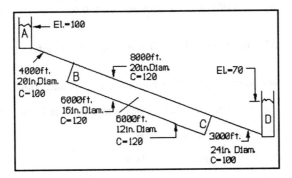

16. The flow through the entire parallel pipe system diagrammed below is horizontal, and is 6.18 cfs. What is the flow through section CD?

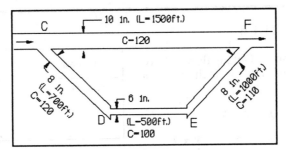

SOLUTIONS

1. **Answer**   48.2 ft.

    The difference in water surface elevations is the HL. Solve for HL in each pipe - add. Use Bernoulli's formula to obtain Ph in tank B.

    Narrow Pipe

    $$Q = .435 \ C \ d^{2.63} \ s^{.54}$$

    $$1 = .435 \ 90 \ .5^{2.63} \ s^{.54}$$

    $$.0329 = s$$

    $$HL = .0329 \ ft./ft. \ x \ 1000 \ ft. = 32.9 \ ft.$$

Wide Pipe

$Q = .435 \ C \ d^{2.63} \ s^{.54}$

$1 = .435 \ 100 \ .67^{2.63} \ s^{.54}$

$.0065 = s$

$HL = .0065$ ft./ft. x 6000 ft. $= 38.9$ ft.

32.9 ft. + 38.9 ft. = 71.8 ft. total HL in system

Set points: point 1 at bottom of tank A and point 2 at bottom of tank B

The difference in velocity heads at a flow of 1 cfs will be negligible.

$Ph_1 + Z_1 = Ph_2 + Z_2 + HL$

$20 + 100 = Ph_2 + 0 + 71.8$

48.2 ft. $= Ph_2$

This is the depth of water in tank B.

---

2. **Answer**   11.1 ft.

Draw Diagram. Obtain flow. Solve for slope, and HL in each segment; add.

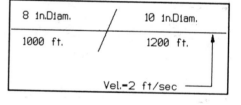

$Q = A \ V$

$Q = .785 \times .83^2 \times 2$

$Q = 1.1$ cfs

8 inch

$Q = .435 \ C \ d^{2.63} \ s^{.54}$

$1.1 = .435 \ 100 \ .67^{2.63} \ s^{.54}$

$.0078 = s$

$HL = .0078$ ft./ft. x 1000 ft. $= 7.8$ ft.

10 inch

$Q = .435 \ C \ d^{2.63} \ s^{.54}$

$1.1 = .435 \ 100 \ .83^{2.63} \ s^{.54}$

$.0027 = s$

HL = .0027 ft./ft. x 1200 ft. = 3.28 ft.

Adding: 7.8 ft. + 3.3 ft. = 11.1 ft. HL in whole system.

3. **Answer**   66.6 psi

Draw Diagram. Considered is the pressure
change between two points, A & E; solve
for elevation of E; find length of pipe A-
E; solve for HL A-E; Use Bernoulli's
formula to find pressure at E.

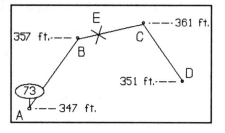

Pipe BC slopes up 4 ft./320 ft.

This equals 1.25 ft./100 ft. - to E.

Point E is at elevation 358.25 ft.

A-B (450 ft. of pipe)

$Q = .435 \ C \ d^{2.63} \ s^{.54}$

$.5 = .435 \ \ 100 \ \ .5^{2.63} \ s^{.54}$

$.0075 = s$

HL = .0075 ft./ft. x 450 ft. = 3.37 ft.

B-E (100 ft. of pipe)

$Q = .435 \ C \ d^{2.63} \ s^{.54}$

$.5 = .435 \ \ 100 \ \ .67^{2.63} \ s^{.54}$

$.0018 = s$

HL = .0018 ft./ft. x 100 ft. = .18 ft.

Adding: 3.37 ft. + .18 ft. = 3.55 ft. total HL

Vh difference insignificant - ignore.

$Ph_A + Z_A = Ph_E + Z_E + HL$

$168.63 + 347 = Ph_E + 358.25 + 3.55$

$153.83 \text{ ft.} = Ph_E$

66.6 psi = pressure

4. **Answer**   A. 127500 ft.
                 B. 1888 ft.

Draw Diagram.

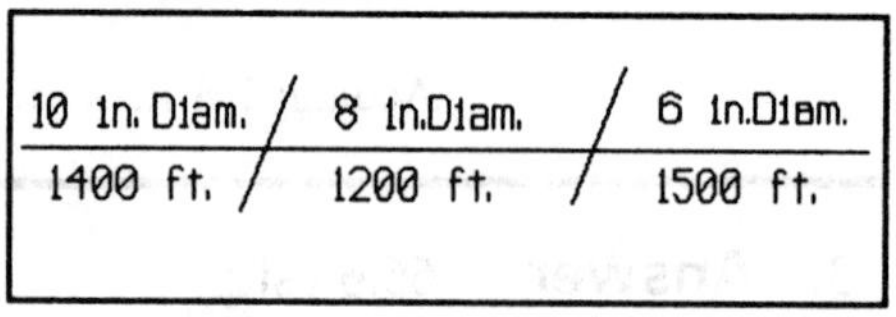

10 in. Diam.	8 in.Diam.	6 in.Diam.
1400 ft.	1200 ft.	1500 ft.

A. Assume a flow; solve for s & HL
    in each pipe. Add. Choose equiv.
    pipe; solve for slope, then L.

Assume Q = 1 cfs

10 inch

$Q = .435 \ C \ d^{2.63} \ s^{.54}$

$1 = .435 \ 100 \ .83^{2.63} \ s^{.54}$

$.0023 = s$

HL = .0023 ft./ft. x 1400 ft. = 3.2 ft.

8 inch

$Q = .435 \ C \ d^{2.63} \ s^{.54}$

$1 = .435 \ 100 \ .67^{2.63} \ s^{.54}$

$.0065 = s$

HL = .0065 ft./ft. x 1200 ft. = 7.8 ft.

6 inch

$Q = .435 \ C \ d^{2.63} \ s^{.54}$

$1 = .435 \ 100 \ .5^{2.63} \ s^{.54}$

$.0270 = s$

HL = .0270 ft./ft. x 1500 ft. = 40 ft.

Adding: 3.2 ft. + 7.8 ft. + 40 ft. = 51 ft. total HL

Equivalent 14 inch pipe; C=100

$Q = .435 \ C \ d^{2.63} \ s^{.54}$

$1 = .435 \ 100 \ 1.167^{2.63} \ s^{.54}$

$.0004 = s$

$$s = \frac{HL}{L}$$

$$.0004 = \frac{51}{L}$$

$$127500 \text{ ft.} = L$$

B.   **Equivalent 6 inch pipe; C=100**

Slope already calculated at Q = 1 cfs; s = .0270.

$$s = \frac{HL}{L}$$

$$.0270 = \frac{51}{L}$$

$$L = 1888 \text{ ft.}$$

---

5. **Answer**   8.6 ft.

Draw Diagram. Newly constructed system; change MGD to cfs; solve for HL in each section; add.

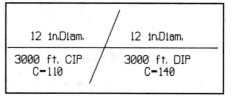

1 MGD x 1.55 cfs/MGD = 1.55 cfs

Old

$$Q = .435 \ C \ d^{2.63} \ s^{.54}$$

$$1.55 = .435 \ \ 110 \ \ 1 \ \ s^{.54}$$

$$.0017 = s$$

HL = .0017 ft./ft. x 3000 ft. = 5.23 ft.

New

$$Q = .435 \ C \ d^{2.63} \ s^{.54}$$

$$1.55 = .435 \ \ 140 \ \ 1 \ \ s^{.54}$$

$$.0012 = s$$

HL = .0012 ft./ft. x 3000 ft. = 3.35 ft.

Adding: 5.23 ft. + 3.35 ft. = 8.58 ft. Total HL in system.

6. **Answer**   A. 93 ft.
                  B. 70.3 ft.
                  C. 70.3 ft.

A. Entire flow goes through bottom loop; solve for s, HL.

$$Q = .435 \ C \ d^{2.63} \ s^{.54}$$

$$10 = .435 \ 120 \ .83^{2.63} \ s^{.54}$$

$$.1162 = s$$

$$HL = .1162 \text{ ft./ft.} \times 800 \text{ ft.} = 93 \text{ ft.}$$

B. Flow splits; assume HL; solve for slope, then flow in each; add flows for total flow; get % flow split; apply to real flow; using real flow, solve for slope, HL in bottom loop.

Assume HL = 10

$$s = \frac{HL}{L} = \frac{10}{1200} = .0083 \text{ ft./ft.}$$

6 inch

$$Q = .435 \ C \ d^{2.63} \ s^{.54}$$

$$Q = .435 \ 90 \ .5^{2.63} \ .0083^{.54}$$

$$Q = .48 \text{ cfs}$$

$$s = \frac{HL}{L} = \frac{10}{800} = .0125$$

10 inch

$$Q = .435 \ C \ d^{2.63} \ s^{.54}$$

$$Q = .435 \ 120 \ .83^{2.63} \ .0125^{.54}$$

$$Q = 3.0 \text{ cfs}$$

Adding flows: .48 cfs + 3.0 cfs = 3.48 cfs total flow (from assumed HL)

$$\frac{.48 \text{ cfs}}{3.48 \text{ cfs}} = .14 \quad \text{Flow through top loop is 14\% of total flow.}$$

$$\frac{3 \text{ cfs}}{3.48 \text{ cfs}} = .86 \quad \text{Flow through bottom loop is 86\% of total flow.}$$

10 cfs (actual total flow) x .14 = 1.4 cfs (actual flow through top loop)

10 cfs (actual total flow) x .86 = 8.6 cfs (actual flow through bottom loop)

Solve for actual HL in bottom loop.

$$Q = .435 \ C \ d^{2.63} \ s^{.54}$$

$$8.6 = .435 \ 120 \ .83^{2.63} \ s^{.54}$$

$$.0879 = s$$

$$HL = .0879 \text{ ft./ft. x } 800 \text{ ft.} = 70.3 \text{ ft.}$$

C. HL across the entire system should be the same. We will work it out to prove it.

Convert to equivalent pipe for entire system; in section b we assumed a HL of 10 ft., and produced a total flow based on this of 3.48 cfs. Now with this flow solve for slope in equivalent pipe, then solve for length of it; then put actual flow through this pipe to get HL.

Equivalent pipe: Choose 12 inch diam. pipe C=100

$$Q = .435 \ C \ d^{2.63} \ s^{.54}$$

$$3.48 = .435 \ 100 \ 1 \ s^{.54}$$

$$.0093 = s$$

$$s = \frac{HL}{L}$$

$$.0093 = \frac{10}{L}$$

$$1075 \text{ ft.} = L \text{ (of equiv. pipe)}$$

$$Q = .435 \ C \ d^{2.63} \ s^{.54}$$

$$10 = .435 \ 100 \ 1 \ s^{.54}$$

$$.0657 = s$$

$$HL = .0657 \text{ ft./ft. x } 1075 \text{ ft.} = 70.5 \text{ ft.} \quad \text{Right On!}$$

---

7. **Answer**   8.4 ft.

Draw Diagram. Assume HL; solve for flow in each; add up; obtain % flow split; multiply by actual flow (given-total) to obtain flow through that section; solve for HL.

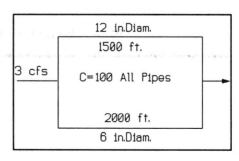

Assume HL = 20 ft.

$$s = \frac{HL}{L} = \frac{20}{1500} = .0133 \text{ ft./ft.}$$

12 inch

$$Q = .435 \ C \ d^{2.63} \ s^{.54}$$

$$Q = .435 \ 100 \ 1 \ .0133^{.54}$$

$$Q = 4.2 \text{ cfs}$$

$$s = \frac{HL}{L} = \frac{20}{2000} = .01 \text{ ft./ft.}$$

6 inch

$$Q = .435 \ C \ d^{2.63} \ s^{.54}$$

$$Q = .435 \ 100 \ .5^{2.63} \ .01^{.54}$$

$$Q = .6 \text{ cfs}$$

Adding flows: 4.2 cfs + .6 cfs = 4.8 cfs Total flow (from assumed HL)

$$\frac{4.2 \text{ cfs}}{4.8 \text{ cfs}} = .88 \qquad \text{(top flow is 88\% of total flow)}$$

$$\frac{.6 \text{ cfs}}{4.8 \text{ cfs}} = .12 \qquad \text{(bottom flow is 12\% of total flow)}$$

Apply top % to actual flow:

3 cfs x .88 = 2.64 cfs

Top

$$Q = .435 \ C \ d^{2.63} \ s^{.54}$$

$$2.64 = .435 \ 100 \ 1 \ s^{.54}$$

$$.0056 = s$$

HL = .0056 ft./ft. x 1500 ft. = 8.4 ft.

8. **Answer**   16.8 ft./sec.

Solve for flow through bottom section; subtract from total flow to get flow through top; solve for velocity in top.

$$Q = A \quad V$$

$$Q = .785 \times .67^2 \times 2$$

$$Q = .7 \text{ cfs (bottom)}$$

4 cfs total flow - .7 cfs bottom flow = 3.3 cfs top

$$Q = A \quad V$$

$$3.3 = .785 \times .5^2 \quad V$$

$$16.8 \text{ ft./sec.} = V$$

---

9. **Answer**   6 inches

Draw Diagram. Convert psi to ft. Solve for s & HL through 6 inch pipe. Subtract from total HL for HL in unknown. Solve for s, then diam.

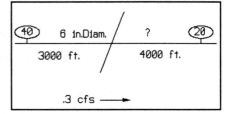

20 psi x 2.31 ft./psi = 46.2 ft.
total HL in system

6 inch

$$Q = .435 \quad C \quad d^{2.63} \quad s^{.54}$$

$$.3 = .435 \quad 80 \quad .5^{2.63} \quad s^{.54}$$

$$.0044 = s$$

HL = .0044 ft./ft. x 3000 ft. = 13.2 ft. HL in 6 inch section.

46.2 ft. Tot. HL in system - 13.2 ft. HL in 6 inch = 33.0 ft. HL in unknown.

$$s = \frac{HL}{L} = \frac{33}{4000} = .0083 \text{ ft./ft. slope of unknown}$$

Unknown

$Q = .435 \ C \ d^{2.63} \ s^{.54}$

$.3 = .435 \quad 80 \quad d^{2.63} \quad .0083^{.54}$

$.44 \ \text{ft.} = d$

Assume 6 inch diameter pipe.

---

10. **Answer**   60.9 cfs

Draw Diagram. Elevation difference =
the total HL. Assume a flow; solve for
HL in each pipe of actual system; add
HL's; Create equivalent pipe; with
assumed flow, solve for slope; using
this and calculated total HL, solve for
length of equiv. pipe. Insert again with
actual head loss, solving for slope;
solve for flow through equiv. pipe.
This is the actual flow through the
system.

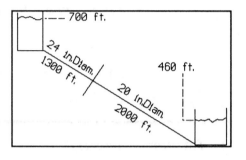

700 ft. - 460 ft. = 240 ft. = HL in system.

Assume flow = 22 cfs

24 inch

$Q = .435 \ C \ d^{2.63} \ s^{.54}$

$22 = .435 \quad 130 \quad 2^{2.63} \quad s^{.54}$

$.006 = s$

$HL = .006 \ \text{ft./ft. x} \ 1300 = 7.8 \ \text{ft.}$

20 inch

$Q = .435 \ C \ d^{2.63} \ s^{.54}$

$22 = .435 \quad 130 \quad 1.67^{2.63} \quad s^{.54}$

$.0143 = s$

$HL = .0143 \ \text{ft./ft. x} \ 2000 \ \text{ft.} = 28.6 \ \text{ft.}$

Adding HL's: 7.8 ft. + 28.6 ft. = 36.4 ft. total HL (from assumed flow)

Create equivalent pipe: 12 inch diam. C=100; Solve for slope.

$$Q = .435 \; C \; d^{2.63} \; s^{.54}$$

$$22 = .435 \; 100 \; 1 \; s^{.54}$$

$$.283 = s$$

$$s = \frac{HL}{L}$$

$$.283 = \frac{36.4}{L}$$

$$128.6 = L \quad \text{(equiv. pipe length)}$$

$$s = \frac{HL}{L} = \frac{240}{128.6} = 1.866 \text{ ft./ft.}$$

$$Q = .435 \; C \; d^{2.63} \; s^{.54}$$

$$Q = .435 \; 100 \; 1 \; 1.866^{.54}$$

$$Q = 60.9 \text{ cfs}$$

---

11. **Answer**   System B

Assume any flow; calculate for total HL in each system; the one with the greater potential HL, will carry less flow.

A. **System A**: Choose a flow; solve for HL in each segment; add up.

$$Q = 2 \text{ cfs}$$

16 inch

$$Q = .435 \; C \; d^{2.63} \; s^{.54}$$

$$2 = .435 \; 120 \; 1.33^{2.63} \; s^{.54}$$

$$.0006 = s$$

$$HL = .0006 \text{ ft./ft.} \times 9000 \text{ ft.} = 5.3 \text{ ft.}$$

12 inch

$$Q = .435 \; C \; d^{2.63} \; s^{.54}$$

$$2 = .435 \; 120 \; 1 \; s^{.54}$$

$$.0024 = s$$

$$HL = .0024 \text{ ft./ft.} \times 6000 \text{ ft.} = 14.3 \text{ ft.}$$

10 inch

$$Q = .435 \ C \ d^{2.63} \ s^{.54}$$

$$2 = .435 \ \ 120 \ \ .83^{2.63} \ \ s^{.54}$$

$$.0059 = s$$

$$HL = .0059 \ ft./ft. \times 3000 \ ft. = 17.7 \ ft.$$

Adding: 5.3 ft. + 14.3 ft. + 17.7 ft. = 37.3 ft. total HL

B. **System B:** Work on parallel segment first, assume a HL; solve for flow in top and bottom segments; add flows; choose equivalent pipe; using new total flow, solve for slope, then length. Then treat as a series pipe system with equivalent pipe in middle.

Parallel segment: assume HL = 10 ft.

Top

$$s = \frac{HL}{L} = \frac{10}{5000} = .002 \ ft./ft.$$

$$Q = .435 \ C \ d^{2.63} \ s^{.54}$$

$$Q = .435 \ \ 120 \ \ .67^{2.63} \ \ .002^{.54}$$

$$Q = .635 \ cfs$$

Bottom

$$s = \frac{HL}{L} = \frac{10}{7000} = .0014$$

$$Q = .435 \ C \ d^{2.63} \ s^{.54}$$

$$Q = .435 \ \ 120 \ \ .83^{2.63} \ \ .0014^{.54}$$

$$Q = .92 \ cfs$$

Adding: (based on assumed HL)

.635 cfs + .92 cfs = 1.56 cfs total flow (from assumed HL)

Equivalent pipe:

$$Q = .435 \ C \ d^{2.63} \ s^{.54}$$

$$1.56 = .435 \ \ 120 \ \ .83^{2.63} \ \ s^{.54}$$

$$.0037 = s$$

Solve for Length.

$$s = \frac{HL}{L}$$

$$.0037 = \frac{10}{L}$$

$$2703 \text{ ft.} = L$$

Now there are three series segments; solve for HL in each, then add.

18 inch

$$Q = .435 \ C \ d^{2.63} \ s^{.54}$$

$$2 = .435 \ 120 \ 1.5^{2.63} \ s^{.54}$$

$$.0003 = s$$

$$HL = .0003 \text{ ft./ft.} \times 1100 \text{ ft.} = 3.3 \text{ ft.}$$

Equiv.

$$Q = .435 \ C \ d^{2.63} \ s^{.54}$$

$$2 = .435 \ 120 \ .83^{2.63} \ s^{.54}$$

$$.0059 = s$$

$$HL = .0059 \text{ ft./ft.} \times 2703 \text{ ft.} = 15.9 \text{ ft.}$$

10 inch

$$Q = .435 \ C \ d^{2.63} \ s^{.54}$$

$$2 = .435 \ 120 \ .83^{2.63} \ s^{.54}$$

$$.0059 = s$$

$$HL = .0059 \text{ ft./ft.} \times 2500 \text{ ft.} = 14.8 \text{ ft.}$$

Adding: 3.3 ft. + 15.9 ft. + 14.8 ft. = 34 ft. total HL

---

12. **Answer**   17.9 ft.

Elevation difference = total HL in pipe system.
Section 3: velocity given - solve for flow.
Section 2 & 3: find slope & HL; add for total HL bottom.
Section 4: solve for HL; find flow.
Add flows in section 4 and 2-3; this gives incoming or outgoing flow.

Section 5: find slope and HL
Add HL's section 1,4,5 for total HL which equals the elevation difference.

$$Q = A \quad V$$

$$Q = .785 \text{ x } .83^2 \text{ x } 2$$

$$Q = 1.1$$

Section 3

$$Q = .435 \quad C \quad d^{2.63} \quad s^{.54}$$

$$1.1 = .435 \quad 100 \quad .83^{2.63} \quad s^{.54}$$

$$.0027 = s$$

$$HL = .0027 \text{ ft./ft. x } 1500 \text{ ft.} = 4.1 \text{ ft.}$$

Section 2

$$Q = .435 \quad C \quad d^{2.63} \quad s^{.54}$$

$$1.1 = .435 \quad 100 \quad .67^{2.63} \quad s^{.54}$$

$$.0078 = s$$

$$HL = .0078 \text{ ft./ft. x } 800 \text{ ft.} = 6.2 \text{ ft.}$$

Adding HL's: 4.1 ft. + 6.2 ft. = 10.3 ft. total HL bottom

Therefore: 10.3 ft. is HL top also.

Section 4

$$s = \frac{HL}{L} = \frac{10.3}{1000} = .0103 \text{ ft./ft.}$$

$$Q = .435 \quad C \quad d^{2.63} \quad s^{.54}$$

$$Q = .435 \quad 100 \quad .83^{2.63} \quad .0103^{.54}$$

$$Q = 2.2 \text{ cfs}$$

Adding flows: 1.1 cfs + 2.2 cfs = 3.3 cfs flow in 1 & 5

Section 1 & 5

Combine length 1 & 5 = 900 ft.

$$Q = .435 \ C \ d^{2.63} \ s^{.54}$$

$$3.3 = .435 \ 100 \ 1 \ s^{.54}$$

$$.0084 = s$$

$$HL = .0084 \ \text{ft./ft.} \times 900 \ \text{ft.} = 7.6 \ \text{ft.}$$

Add HL's: Top (4) 10.3 ft. + Ends (1 & 5) 7.6 ft. = 17.9 ft.

---

13. **Answer**   A. 2950 ft.
               B. 91.8 ft.

Draw Diagram.

	4 in.Diam.	570 ft.
	C-140	
3000 gpm	6 in.Diam.	900 ft.
	C-80	
	8 in.Diam.	1200 ft.
	C-100	

A. Assume a HL; obtain slope, flow for each pipe in actual system; add flows for total flow; create equivalent pipe; using total flow from assumed HL, solve for slope; with assumed HL & slope of equiv. pipe, solve for length.

Assume HL = 100 ft.

4 inch

$$s = \frac{HL}{L} = \frac{100}{570} = .1754 \ \text{ft.}$$

$$Q = .435 \ C \ d^{2.63} \ s^{.54}$$

$$Q = .435 \ 140 \ .33^{2.63} \ .1754^{.54}$$

$$Q = 1.3 \ \text{cfs}$$

6 inch

$$s = \frac{HL}{L} = \frac{100}{900} = .1111 \ \text{ft.}$$

$$Q = .435 \ C \ d^{2.63} \ s^{.54}$$

$$Q = .435 \ 80 \ .5^{2.63} \ .1111^{.54}$$

$$Q = 1.7 \ \text{cfs}$$

8 inch

$$s = \frac{HL}{L} = \frac{100}{1200} = .0833 \ \text{ft.}$$

$$Q = .435 \ C \ d^{2.63} \ s^{.54}$$

$$Q = .435 \ 100 \ .67^{2.63} \ .0833^{.54}$$

$$Q = 4.0 \ cfs$$

Adding flows: (based on assumed HL)

1.3 cfs + 1.7 cfs + 4.0 cfs = 7.0 cfs Total flow (from assmed HL)

Choose equivalent pipe: 12 inch diam. C=100

$$Q = .435 \ C \ d^{2.63} \ s^{.54}$$

$$7 = .435 \ 100 \ 1 \ s^{.54}$$

$$.0339 = s$$

Solve for length of equivalent pipe.

$$s = \frac{HL}{L}$$

$$.0339 = \frac{100}{L}$$

$$2950 \ ft. = L$$

B. Use actual flow to obtain HL of equivalent pipe (which is the same as the HL of our system).

$$\frac{3000 \ gpm}{7.48 \times 60} = 6.68 \ cfs$$

$$Q = .435 \ C \ d^{2.63} \ s^{.54}$$

$$6.68 = .435 \ 100 \ 1 \ s^{.54}$$

$$.0311 = s$$

HL = .0311 ft./ft. x 2950 ft. = 91.8 ft.

---

14. **Answer**   A. 5061 ft.
                   B. 6.75 cfs

Draw Diagram. Assume Q = 1.

20 in.Diam.	16 in.Diam.	12 in.Diam.
10,000 ft.	8000 ft.	L- ?
C-120	C-120	C-120

A. Solve for HL in 15 inch pipe. Total HL in actual system is the same. Using assumed flow, obtain HL for each of the first two pipes. Difference between HL (equiv. pipe) and HL in actual system just calculated, is HL in the 12 inch pipe - at that flow. Solve for length of 12 inch pipe. This will be the correct pipe length to yield a HL that makes the actual system equiv. to the equiv. pipe.

Equiv.

$Q = .435 \ C \ d^{2.63} \ s^{.54}$

$1 = .435 \ 100 \ 1.25^{2.63} \ s^{.54}$

$.0003 = s$

HL = .0003 ft./ft. x 16500 ft. = 5.14 ft.

20 inch

$Q = .435 \ C \ d^{2.63} \ s^{.54}$

$Q = .435 \ 120 \ 1.67^{2.63} \ s^{.54}$

$.00005 = s$

HL = .00005 ft./ft. x 10000 ft. = .5 ft.

16 inch

$Q = .435 \ C \ d^{2.63} \ s^{.54}$

$Q = .435 \ 120 \ 1.33^{2.63} \ s^{.54}$

$.00016 = s$

HL = .00016 ft./ft. x 8000 ft. = 1.3 ft.

Adding:
.5 ft. + 1.3 ft. = 1.8 ft. HL in AB & BC of actual system at Q=1

5.14 ft. (HL Equiv. pipe at Q=1) - 1.8 ft. = 3.34 ft. (HL CD at Q=1)

12 inch

$Q = .435 \ C \ d^{2.63} \ s^{.54}$

$1 = .435 \ 120 \ 1 \ s^{.54}$

$.00066 = s$

$$s = \frac{HL}{L}$$

$$.00066 = \frac{3.34}{L}$$

5061 ft. = L

B. Tot. HL is given; to solve for flow we must use the equivalent pipe method. We have already worked through the procedure in section A. Use the same figures. Assume a flow (1); solve for HL in each pipe of actual system; add, solve for slope in equiv. pipe (use the same one); find a new length for it; Using s=HL/L, insert new L and 135 ft. for HL, and get s; solve for Q. This is the flow through the actual system.

Assume Q = 1 cfs.

From section A:

HL in 20 inch pipe = .5 ft.

HL in 16 inch pipe = 1.3 ft.

s for 12 inch pipe = .00066

HL (12 inch) = .00066 ft./ft. x 3000 ft. = 1.98 ft.

Adding: .5 ft. + 1.3 ft. + 1.98 ft. = 3.78 ft. total HL at Q=1

From section A:

s for equiv. pipe = .0003 ft./ft.

Solve for length:

$$s = \frac{HL}{L}$$

$$.0003 = \frac{3.78}{L}$$

$$12600 \text{ ft.} = L$$

Now get real slope for equiv. pipe.

$$s = \frac{HL}{L} = \frac{135}{12600} = .0107 \text{ ft./ft.}$$

Solve for flow in equiv. pipe.

$$Q = .435 \; C \; d^{2.63} \; s^{.54}$$

$$Q = .435 \; 100 \; 1.25^{2.63} \; .0107^{.54}$$

$$Q = 6.75 \text{ cfs}$$

15. **Answer**   6.7 cfs

HL is given as the difference in elevation between the two reservoirs (30 ft.). Bottom segment of parallel system is composed of two pipes in series. Convert to equivalent pipe: assume flow, find HL in each, add, find slope of equivalent pipe, solve for length. Parallel section is now composed of two pipes; convert to one equiv. pipe: assume HL, find flow in each, add; using total flow, find slope for equivalent; solve for length of equivalent pipe. Entire system now composed of 3 pipes in series; convert to one equivalent pipe: assume flow, find HL in each, add; using assumed flow, find slope of equivalent, solve for length. Now take the actual HL (30 ft.) and solve for s and flow for entire system.

Bottom of Parallel System

Assume Q = 10 cfs   Using actual system:

Actual 16 inch

$Q = .435 \ C \ d^{2.63} \ s^{.54}$

$10 = .435 \ 120 \ 1.33^{2.63} \ s^{.54}$

$.0117 = s$

HL = .0117 ft./ft. x 6000 ft. = 70.1 ft.

12 inch

$Q = .435 \ C \ d^{2.63} \ s^{.54}$

$10 = .435 \ 120 \ 1 \ s^{.54}$

$.0469 = s$

HL = .0469 ft./ft. x 6000 ft. = 281.3 ft.

Adding HL's:
70.1 ft. + 281.3 ft. = 351.4 ft. Total HL (from assumed flow).

Equivalent pipe - choose 12 inch diam. C=120

From above, slope at this flow is .0469.

$$s = \frac{HL}{L}$$

$$.0469 = \frac{351.4}{L}$$

7493 ft. = L

The pipe now becomes the bottom segment of a parallel system. (7493 ft. of 12 inch pipe C=120).

Convert parallel system to equivalent pipe:

Assume HL = 20 ft. Solve for flow.

Actual 20 inch

$$s = \frac{HL}{L} = \frac{20}{3000} = .0025 \text{ ft./ft.}$$

$Q = .435 \ C \ d^{2.63} \ s^{.54}$

$Q = .435 \ 120 \ .67^{2.63} \ .0025^{.54}$

$Q = 7.9 \text{ cfs}$

New Equiv. 12 inch

$$s = \frac{HL}{L} = \frac{20}{7493} = .0027 \text{ ft./ft.}$$

$Q = .435 \ C \ d^{2.63} \ s^{.54}$

$Q = .435 \ 120 \ 1 \ .0027^{.54}$

$Q = 2.14 \text{ cfs}$

Adding flows: 7.9 cfs + 2.14 cfs = 10 cfs

Choose equivalent pipe: 12 inch diam. C=120; solve for slope.

$Q = .435 \ C \ d^{2.63} \ s^{.54}$

$10 = .435 \ 120 \ 1 \ s^{.54}$

$.0469 = s$

$$s = \frac{HL}{L}$$

$.0469 = \frac{20}{L} = 426 \text{ ft. length}$

This pipe now becomes the middle segment of the series system. (426 ft. of 12 inch pipe C=120)

Convert this to equivalent pipe: Assume Q=1 cfs.

20 inch

$Q = .435 \ C \ d^{2.63} \ s^{.54}$

$1 = .435 \ 100 \ 1.67^{2.63} \ s^{.54}$

$.00008 = s$

HL = .00008 ft./ft. x 4000 ft. = .3 ft.

12 inch

$Q = .435 \ C \ d^{2.63} \ s^{.54}$

$1 = .435 \ 120 \ 1 \ s^{.54}$

$.0007 = s$

$HL = .0007 \ ft./ft. \times 426 \ ft. = .3 \ ft.$

24 inch

$Q = .435 \ C \ d^{2.63} \ s^{.54}$

$1 = .435 \ 100 \ 2^{2.63} \ s^{.54}$

$.00003 = s$

$HL = .00003 \ ft./ft. \times 3000 \ ft. = .1 \ ft.$

Adding HL's:
.3 ft. + .3 ft. + .1 ft. = .7 ft. HL (based on assumed Q=1 cfs)

Choose equivalent pipe: 18 inch C=100; solve for slope.

$Q = .435 \ C \ d^{2.63} \ s^{.54}$

$1 = .435 \ 100 \ 1.5^{2.63} \ s^{.54}$

$.0001 = s$

$s = \dfrac{HL}{L}$

$.0001 = \dfrac{.7}{L}$

$7000 \ ft. = L$

The pipe now becomes one: (7000 ft. of 18 inch pipe C=100)

Use actual HL given; solve for slope

$s = \dfrac{HL}{L} = \dfrac{30}{7000} = .0043 \ ft./ft.$

$Q = .435 \ C \ d^{2.63} \ s^{.54}$

$Q = .435 \ 100 \ 1.5^{2.63} \ .0043^{.54}$

$Q = 6.7 \ cfs$

16. **Answer**   1.48 cfs

Bottom section is composed of three pipes in series. Convert to one equivalent pipe: assume flow; solve for HL in actual system; add HL's; create equivalent; find slope; solve for length. Now this is a parallel system; assume HL; solve for flow in each; add; get % split; apply to actual flow for flow in CD.

Convert bottom to equivalent pipe:

Assume Q=1 cfs.

8 inch

$Q = .435 \ C \ d^{2.63} \ s^{.54}$

$1 = .435 \ 120 \ .67^{2.63} \ s^{.54}$

$.0046 = s$

$HL = .0046$ ft./ft. x 700 ft. $= 3.25$ ft.

6 inch

$Q = .435 \ C \ d^{2.63} \ s^{.54}$

$1 = .435 \ 100 \ .5^{2.63} \ s^{.54}$

$.027 = s$

$HL = .027$ ft./ft. x 500 ft. $= 13.5$ ft.

8 inch

$Q = .435 \ C \ d^{2.63} \ s^{.54}$

$1 = .435 \ 110 \ .67^{2.63} \ s^{.54}$

$.0055 = s$

$HL = .0055$ ft./ft. x 1000 ft. $= 5.5$ ft.

Adding HL's: 3.25 ft. + 13.5 ft. + 5.5 ft. = 22.3 ft. total HL

Choose equivalent pipe: 8 inch diam. C=110

From above, slope for this = .0055 ft./ft..

$$s = \frac{HL}{L}$$

$$.0055 = \frac{22.3}{L}$$

$$4049 \text{ ft.} = L$$

Now a parallel system remains; solve for flows.

Assume an HL: (50 ft.)

Top

$$s = \frac{HL}{L} = \frac{50}{1500} = .0333 \text{ ft./ft.}$$

$$Q = .435 \ C \ d^{2.63} \ s^{.54}$$

$$Q = .435 \ 130 \ .83^{2.63} \ .0333^{.54}$$

$$Q = 5.1 \text{ cfs}$$

Bottom

$$s = \frac{HL}{L} = \frac{50}{4049} = .0124 \text{ ft./ft.}$$

$$Q = .435 \ C \ d^{2.63} \ s^{.54}$$

$$Q = .435 \ 110 \ .67^{2.63} \ .0124^{.54}$$

$$Q = 1.6 \text{ cfs}$$

Adding flows:
5.1 cfs + 1.6 cfs = 6.7 cfs Total flow (from assumed HL)

$$\frac{5.1}{6.7} = .76 \quad \text{(flow through top is 76\% of total flow)}$$

$$\frac{1.6}{6.7} = .24 \quad \text{(flow through bottom is 24\% of total flow)}$$

Bottom section carries 24% of total flow.

6.18 cfs x .24 = 1.48 cfs

CHAPTER 9

MINOR HEAD LOSS

In Chapter 7 it was stated briefly that friction head loss can be separated into two types: Major Head Loss (from friction encountered along the length of the pipe) and Minor Head Loss (pressure loss at bends, valves, fittings, and changes in pipe diameter). Extra turbulence is created at these areas, and the resultant head loss is often significant enough to be included in calculations. It is almost always less than major head loss, however, and its inclusion is optional. For example, if an 8 inch diameter pipe carries water for two miles, and has only one 90 degree elbow at mid-length, the minor loss from this is considered too small to count. But inside a factory, where the piping turns, enlarges, contracts and encounters valves every few feet, the minor loss would be significant. A city water distribution system has valves and laterals at frequent intervals, and an approximation of minor loss would have to be made to determine its significance. A general guideline which can be used: if the minor head loss is less than five percent of the total head loss, ignore it.

EQUIVALENT PIPE TECHNIQUE

Determination of minor head loss is simple and straightforward, and a standard nomograph in common use is included at the end of the chapter. For calculations, each fitting is converted to a length of straight pipe of the same diameter which would cause the same head loss. This makes further calculation easy. Once the fitting has merely been converted to an extra length of pipe, calculations for the total length will include both major and minor losses.

Inspecting the nomograph, pipe size is given as both Inside Diameter and Nominal Diameter. Nominal Diameter is just a designation of pipe size. For instance, we call it a six inch pipe (Nominal Diameter), although the actual inside diameter of the pipe is not exaclty six inches. For practical use with a ruler here, however, the difference is insignificant.

Note the difference in equivalent pipe length between a fully open gate valve, and a 3/4 closed gate valve. In pumping operations, it is common practice to control flow rate by partially closing a valve on the pump discharge, thereby increasing the minor head loss, and restricting the flow. Using another example, a pressure regulating valve senses a pressure higher than its setpoint on the downstream side of the valve, and closes slightly to increase the minor head loss, thus dropping the pressure. In water distribution systems, we can readily understand the importance of proper valve maintanance and record keeping. A shutoff valve in a large main left half closed by accident would significantly affect the line pressure.

Larger diameter fittings yield larger equivalent lengths of straight pipe. This seems to contradict our previous calculations for friction head loss (small pipes have more head loss), but the attempt with this nomograph is to keep the head loss value for a particular fitting the same no matter what the size. After converting that fitting to an equivalent length of pipe, head loss will then be calculated based on the diameter, and that same type of fitting would end up yielding a calculated smaller head loss if it were on a larger pipe, unless the nomograph had designated larger equivalent pipe length for larger pipes.

Note the diagram shown for "Ordinary Entrance". This refers to an entrance <u>to the pipe</u>. Similarly, "Ordinary Exit" would be an exit <u>from the pipe</u>.

The odd configuration called "Borda Entrance" should perhaps have been drawn like this:

The most frequently seen application of a Borda Entrance is at the end of the intake line of a pump drawing a suction lift.

Borda Entrance

For enlargements and contractions, measure with the smaller pipe size on the nomograph. Recognize that two separate hydraulic phenomena are occurring in these areas of pipe enlargements or contractions:

1. Pressure drop because of extra turbulence encountered in passage to a pipe of different diameter (minor head loss). This is a permanent loss of pressure, which must be made up by the source.

2. Pressure and velocity change encountered because the pipe changed diameter (from $Q=AV$ and Bernoulli's Theorem). This change could be only temporary. Original pressure and velocity would return if downstream the pipe returned to the original diameter, and the only pressure drop remaining would be from major and minor head losses through this section.

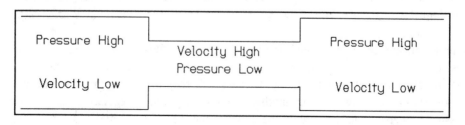

When faced with a fitting not listed on the nomograph - estimate. Use the closest one. Use care with reading nomographs; misplacing the ruler by a fraction of an inch may change the reading by a great deal.

VELOCITY HEAD TECHNIQUE

There are a few minor losses which may be expressed in terms of Velocity Head ($V^2/2g$). These have been worked out by experimentation and measurement, and are useful in calculations where Bernoulli's Theorem is in use, and a Velocity Head value is readily at hand. This method directly converts to a head loss value for the fixture:

Ordinary Exit:
Head Loss = 1 Velocity Head

This makes sense. The water loses its energy due to velocity upon entrance to the tank, because it loses all its velocity.

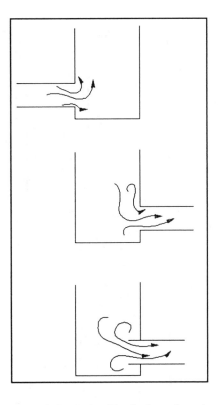

Ordinary Entrance:
Head Loss = ½ Velocity Head

Turbulence is created in getting the water out and around the 90 degree angle and into the pipe.

Borda Entrance:
Head Loss = 1 Velocity Head

Extra turbulence is created in getting the water out and around the 180 degree angle and into the pipe.

Equating the two techniques, Ordinary Exit is entered into the Equivalent length nomograph at the same place as Borda Entrance.

Most commonly, conversion to equivalent pipe length is the method of choice. Calculations from previous chapters remain the same when minor head loss is included. The only difference is that with this inclusion, the pipe length has been extended.

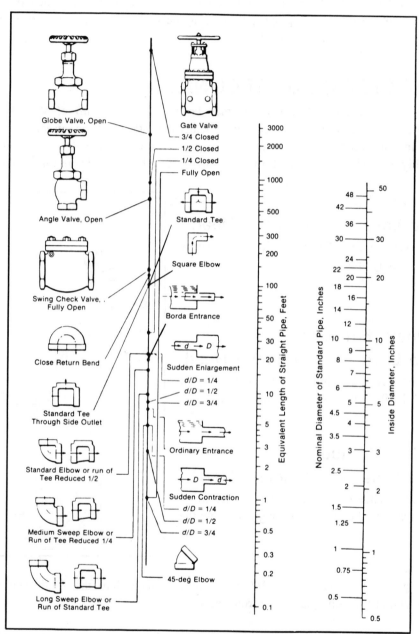

Reprinted with permission of Crane Co.

Resistance to Flow from Valves and Fittings

PROBLEMS

Minor Loss is to be included in all calculations.

1. Find the equivalent length of pipe for each of the following:
    A. 45 degree elbow - 10 inch diameter
    B. 4 inch diameter Borda Entrance
    C. Sudden enlargement: 4 inches to 8 inches
    D. Ordinary Exit: 36 inches
    E. 1 inch diameter close return bend

2. A 6 inch diameter main (C=110) is carrying a flow of 1 cfs. What head loss is produced by two regular flanged 45 degree elbows and an open globe valve?

3. Two hundred feet of 12 inch diameter pipe with water flowing has a slope of .02 ft./ft. There is a 3/4 closed gate valve on the pipe.
    A. What is the head loss in the pipe?
    B. What would be the head loss if the valve were fully open?

4. If all pipe is 6 inch diameter, 10 year old cast iron, and Q=3 cfs, find the total head loss through this system.

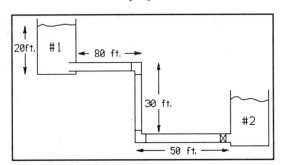

5. Water entering a house through a 1 inch diameter horizontal copper pipe makes two 90 degree turns before the pipe ends at a faucet. If the pressure at entrance is 30 psi, and the pipe is 25 ft. long, when the faucet is fully open, what is the flow?

6. Water travels from point 1 to point 2 through 1000 ft. of 10 yr. old cast iron pipe, 12 inches in diameter. Installed in the line are one swing check valve, one open globe valve, and two square elbows. If the flow is steady at 2 cfs, how much pressure does the water lose in traveling from point 1 to point 2 (psi)?

7. A tank discharges water through a 10 inch diameter pipe, 900 ft. long (C=120), at a velocity of 5 ft./sec. What is the head loss through the system:
    A. by the Velocity Head technique
    B. by the Equivalent Pipe technique

8. A 24 inch diameter transmission main carries 7 cfs for a length of two miles. Along the pipe are three 45 degree elbows and 4 open gate valves. What percentage of the total head loss is caused by the fittings on this pipe?

9. Diagrammed to the right is a heat exchanger. What is the flow (cfs) through this unit? (All pipe is 4 inch diameter copper pipe).

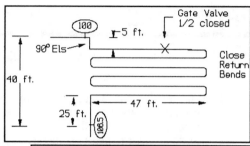

10. What is the elevation of the water surface in tank #2? All valves are fully open; C=110; all pipe is 12 inch diameter; Q=800 gpm.

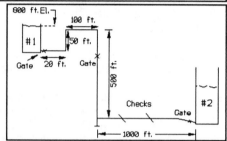

11. A pump drawing a 20 ft. suction lift delivers 200 gpm. What is the pressure drop in the 6 inch diameter pump suction line, which is 10 year old CIP, and includes a standard elbow and a foot valve?

12. Determine the head loss in the system. All bends are standard elbows. Q=3.2 cfs.

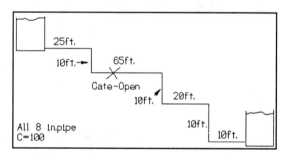

13. Water service was extended in a galvanized iron loop to supply a temporary establishment. During pipeline construction, a shutoff valve on the line was inadvertently left half closed, as shown in the diagram. What is the flow through the loop?

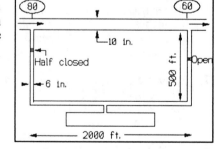

14. An 8 inch diameter pipe (C=100) enlarges to a 12 inch diameter pipe. If the 8 inch pipe is 1200 ft. long, the 12 inch pipe is 800 ft. long (C=140), and the flow is 2 cfs:
    A. What is the head loss in the system?
    B. What is the velocity in the 12 inch diameter pipe?

15. An 8 inch diameter pipe 1000 ft. long takes water from an upper reservoir. On this pipe are a gate valve, 3 standard elbows, and a globe valve. The pipe is connected to a 6 inch diameter pipe 50 ft. long with an angle valve and a standard tee which delivers the water to a lower reservoir. If the difference in the water levels of the two reservoirs is 50 ft., what is the flow through this system ($C = 100$ all pipes)?

SOLUTIONS

1. **Answer**   A.   13 ft. equivalent pipe
                     B.   13 ft. equivalent pipe
                     C.   6.5 ft. equivalent pipe
                     D.   120 ft. equivalent pipe
                     E.   6.5 ft. equivalent pipe

2. **Answer**   4 ft.

   Obtain equivalent length for fittings; add; solve for HL.

globe valve (6 inches)	160 ft.
45 deg. elbow (2) (6 inches)	<u>16 ft.</u>
	176 ft. length

   $$Q = .435 \ C \ d^{2.63} \ s^{.54}$$

   $$1 = .435 \ \ 110 \ \ .5^{2.63} \ \ s^{.54}$$

   $$.0227 = s$$

   $$HL = .0227 \text{ ft./ft. x } 176 \text{ ft.} = 4 \text{ ft. HL created by fittings}$$

3. **Answer**   A.   20 ft.
                     B.   4.1 ft.

   A.   Obtain equivalent length of valve; add to pipe length; solve for HL.

3/4 closed gate valve (12 in.)	800 ft.
actual pipe length	<u>200 ft.</u>
	1000 ft.

   $$s = \frac{HL}{L}$$

   $$.02 = \frac{HL}{1000 \text{ ft.}}$$

   $$20 \text{ ft.} = HL$$

B. Repeat with fully open gate valve.

Fully open gate valve (12 in.)	6.5 ft.
Actual pipe length	200 ft.
	206.5 ft.

206.5 ft. x .02 ft./ft. = 4.1 ft. HL

---

4. **Answer**   40 ft.

Add straight pipe lengths; convert fittings to equiv. pipe; add on; solve for HL.

Straight pipe (80 ft. + 30 ft. + 50 ft.)	160	ft.
Borda entrance (6 in.)	18	ft.
Standard elbows (2) (6 in.)	34	ft.
Gate valve (6 in.)	3.5	ft.
Ordinary exit (6 in.)	18	ft.
	233.5	ft. total pipe length

10 yr. old CIP: C=110

$$Q = .435 \ C \ d^{2.63} \ s^{.54}$$

$$3 = .435 \ 110 \ .5^{2.63} \ s^{.54}$$

$$.1733 = s$$

HL = .1733 ft./ft. x 233.5 ft. = 40 ft.

Note: Tank #1 is only 20 ft. deep. The HL is 40 ft. If the 30 ft. length of pipe is a horizontal pipe, the water would not get to Tank #2 at this flow. In order to maintain this flow, there must be a 20 ft. vertical drop in this pipe.

---

5. **Answer**   .095 cfs

Draw Diagram. Solve for head loss; obtain equivalent length for fittings; add to pipe length; solve for slope, then flow.

New copper pipe (C=140)

Head loss = pressure at entrance (it is all lost at faucet)

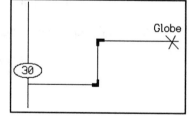

30 psi x 2.31 ft./psi = 69.3 ft. Head Loss

Square elbows (2) (1 in.)	10 ft.
Open globe valve (1) (1 in.)	25 ft.
Actual pipe length	25 ft.
	60 ft.

$$s = \frac{HL}{L} = \frac{69.3}{60} = 1.155$$

$$Q = .435 \ C \ d^{2.63} \ s^{.54}$$

$$Q = .435 \ 140 \ .0833^{2.63} \ 1.155^{.54}$$

$$Q = .095 \ cfs$$

---

6. **Answer**   1.9 psi

Draw Diagram. Find equivalent pipe length for fittings; add to actual pipe length; solve for HL; change to psi.

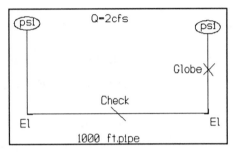

Swing check (12 in.)	80 ft.
Globe valve (12 in.)	325 ft.
Square els (2) (12 in.)	136 ft.
Equiv. length for fittings	541 ft.

541 ft. + 1000 ft. actual pipe length = 1541 ft. total pipe length.

10 yr. old CIP: C=110

$$Q = .435 \ C \ d^{2.63} \ s^{.54}$$

$$2 = .435 \ 110 \ 1 \ s^{.54}$$

$$.0028 = s$$

$$HL = .0028 \ \text{ft./ft.} \times 1541 \ \text{ft.} = 4.3 \ \text{ft.}$$

$$\frac{4.3 \ \text{ft.}}{2.31 \ \text{ft./psi}} = 1.9 \ \text{psi}$$

---

7. **Answer**   A. 9.45 ft.
                       B. 9.43 ft.

Draw Diagram.

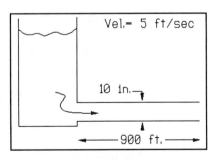

A. Solve for Velocity Head. Solve for flow, then HL using actual length of pipe. Equate Vh to entrance HL; add to HL in straight pipe.

$$Vh = \frac{V^2}{2g} = \frac{5^2}{64.4} = .388 \ \text{ft.}$$

$$Q = A \quad V$$

$$Q = .785 \text{ x } .83^2 \text{ x } 5$$

$$Q = 2.7 \text{ cfs}$$

$$Q = .435 \text{ C } d^{2.63} \text{ s}^{.54}$$

$$2.7 = .435 \quad 120 \quad .83^{2.63} \text{ s}^{.54}$$

$$.0103 = s$$

$$HL = .0103 \text{ ft./ft. x } 900 \text{ ft.} = 9.253 \text{ ft.}$$

Head loss for Ordinary entrance $= \frac{1}{2}$ Vh

HL for this fixture $= .194$ ft.

HL (straight pipe) + HL (fixture) = total HL

$$9.253 + .194 = 9.45 \text{ ft.}$$

B. Obtain equivalent length straight pipe for ordinary entrance; add to pipe length; solve for flow, HL.

Straight pipe (10 in.)	900 ft.
Ordinary entrance (10 in.)	15 ft.
	915 ft. total pipe length

Flow with this fitting, from above = 2.7 cfs.

Slope, from above = .0103 ft./ft.

$$HL = .0103 \text{ ft./ft. x } 915 \text{ ft.} = 9.43 \text{ ft.}$$

8. **Answer**   1.4%

Obtain equivalent length for fittings; add to pipe length for total; divide this into equivalent length for fittings.

Pipe length: 2 miles x 5280 ft./mile = 10,560 ft.

45 degree elbows (3) (24 in.)	90 ft.
Gate valves, open (4) (24 in.)	60 ft.
	150 ft. equiv. length for fittings

150 ft. (fittings) + 10,560 ft. (pipe) = 10,710 ft. total

$$\frac{150}{10710} = .014 = 1.4\%$$

Equivalent length for fittings is 1.4% of total length. Therefore, head loss due to fittings is 1.4% of total head loss. It would be considered negligible.

9. **Answer**   .63 cfs

Using Bernoulli's formula, solve for total HL. Obtain pipe length; obtain equivalent pipe length for fittings; add; solve for slope, then flow.

Top gage: 100 psi x 2.31 ft./psi = 231 ft.

Bottom gage: 106.5 psi x 2.31 ft./psi = 246 ft.

Single diameter pipe: Velocity Heads cancel.

Ph + Z = Ph + Z + HL

231 + 40 = 246 + 0 + HL

25 ft. = HL

Six runs of 47 ft. length + 30 ft. vertical length = 312 ft. pipe.

Standard elbows (3) (4 in.)	33 ft.
Close return bends (5) (4 in.)	125 ft.
½ closed gate valve (4 in.)	65 ft.
Straight pipe	312 ft.
	535 ft. total length

New steel pipe = C=140

$$s = \frac{HL}{L} = \frac{25}{535} = .0467 \text{ ft./ft.}$$

$$Q = .435 \ C \ d^{2.63} \ s^{.54}$$

$$Q = .435 \ \ 140 \ \ .33^{2.63} \ \ .0467^{.54}$$

$$Q = .63 \text{ cfs}$$

---

10. **Answer**   795 ft.

Tank #2 surface elevation will be Tank #1 surface elevation, minus the HL; add up pipe lengths; convert fittings to equivalent pipe; add on. Change flow to cfs; solve for HL; subtract from Tank #1 surface elevation.

Straight pipe (20 ft. + 50 ft. + 100 ft. + 500 ft. + 1000 ft.)	1670 ft.
Gate valves (3) (12 in.)	21 ft.
Ordinary entrance (12 in.)	18 ft.
Square elbows (3) (12 in.)	204 ft.
Tee (12 in.)	70 ft.
Check valves (2) (12 in.)	160 ft.
45 degree elbow (12 in.)	15 ft.
Ordinary exit (12 in.)	37 ft.
	2195 ft. total pipe length

$$\frac{800 \text{ gpm}}{60 \times 7.48} = 1.78 \text{ cfs}$$

$$Q = .435 \ C \ d^{2.63} \ s^{.54}$$

$$1.78 = .435 \ 110 \ 1 \ s^{.54}$$

$$.0023 = s$$

$$HL = .0023 \text{ ft./ft.} \times 2195 \text{ ft.} = 5 \text{ ft.}$$

Surface elev. Tank #1	800 ft.
Head loss	- 5 ft.
	795 ft. Surface elev. Tank #2

---

11. **Answer**   8.9 psi

Draw Diagram.   Obtain equivalent length for fittings; add to pipe length; solve for head loss; include elevation head.

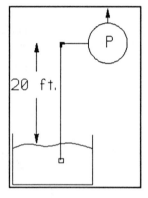

10 year old CIP   (C=110)

Borda entrance (6 in.)	18 ft.
Std. elbow (1) (6 in.)	17 ft.
Foot valve (swing check) (6 in.)	40 ft.
Actual pipe length	20 ft.
	95 ft.

Flow = 200 gpm x 1440 min. day = 288,000 gpd

.288 MGD x 1.55 cfs/MGD = .45 cfs

$$Q = .435 \ C \ d^{2.63} \ s^{.54}$$

$$.45 = .435 \ 110 \ .5^{2.63} \ s^{.54}$$

$$.0052 = s$$

.0052 ft./ft. x 95 ft. = .5 ft. head loss (from friction and fittings)

Including 20 ft. of elevation head loss in the lift, total head loss = 20.5 ft.

$$\frac{20.5 \text{ ft.}}{2.31 \text{ ft./psi}} = 8.9 \text{ psi pressure drop}$$

12. **Answer**   18.1 ft.

Add pipe lengths; convert fittings to straight pipe length; add on; solve for HL.

Straight pipe (25 ft. + 10 ft. + 65 ft. + 10 ft. + 20 ft. + 10 ft. + 10 ft.)	150 ft.
Ordinary entrance (8 in.)	12.5 ft.
Gate valve (8 in.)	4.5 ft.
Standard elbows (6) (8 in.)	132 ft.
Ordinary Exit (8 in.)	25 ft.
	324 ft. total pipe L

$$Q = .435 \ C \ d^{2.63} \ s^{.54}$$

$$3.2 = .435 \ 100 \ .67^{2.63} \ s^{.54}$$

$$.056 = s$$

$$HL = .056 \ ft./ft. \times 324 \ ft. = 18.1 \ ft.$$

---

13. **Answer**   .85 cfs

Obtain equivalent length for fittings; add to pipe length; solve for feet of head loss through top section; will be the same for the loop (parallel pipes); solve for flow through loop.

Std. tee through side (2) (6 in.)	70 ft.
Std. elbows (2) (6 in.)	34 ft.
Open gate (1) (6 in.)	4 ft.
Half closed gate (1) (6 in.)	100 ft.
Actual pipe length	3000 ft.
	3208 ft.

Head loss through top = 80 psi - 60 psi = 20 psi

20 psi x 2.31 ft./psi = 46.2 ft. head loss

Therefore, head loss through bottom = 46.2 ft.

$$s = \frac{HL}{L} = \frac{46.2}{3208} = .0144 \ ft./ft.$$

$$Q = .435 \ C \ d^{2.63} \ s^{.54}$$

$$Q = .435 \ 120 \ .5^{2.63} \ .0144^{.54}$$

$$Q = .85 \ cfs$$

14. **Answer**   A. 29.6 ft.
                 B. 2.55 ft./sec.

Draw Diagram.

A. Solve for HL in each pipe; add;
    only fitting is sudden enlargement
    (about 13 ft. of equiv. pipe length -
    ignore it.)

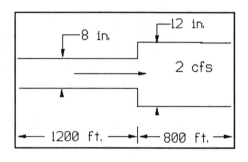

$$Q = .435 \ C \ d^{2.63} \ s^{.54}$$

$$2 = .435 \ \ 100 \ \ .67^{2.63} \ s^{.54}$$

$$.0235 = s$$

HL = .0235 ft./ft. x 1200 ft. = 28.2 ft.

$$Q = .435 \ C \ d^{2.63} \ s^{.54}$$

$$2 = .435 \ \ 140 \ \ 1 \ s^{.54}$$

$$.0018 = s$$

HL = .0018 ft./ft. x 800 ft. = 1.4 ft.

Adding HL's: 28.2 ft. + 1.4 ft. = 29.6 ft. total HL

B. $Q = A \ V$

$$2 = .785 \times 1 \ V$$

$$2.55 \ ft./sec. = V$$

---

15. **Answer**   2.1 cfs

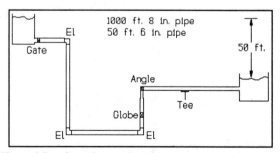

Draw Diagram. Convert
fittings to lengths of equivalent
pipe; add to connecting straight
pipe; solve for length of a pipe
equivalent to the system
(assume Q, solve for HL in
each pipe, add HL's; create
equiv. pipe, solve for slope,
solve for length). Using given HL and L of equiv., solve for slope, then flow.

8 inch pipe

Straight pipe	1000 ft.
Ordinary entrance	12 ft.
Gate Valve	5 ft.
Elbows (3)	66 ft.
Globe valve	200 ft.
	1303 ft.

6 inch pipe

Straight pipe	50 ft.
Sudden contraction	5 ft.
Angle valve	90 ft.
Tee	12 ft.
Ordinary exit	18 ft.
	175 ft.

Assume Q = 3 cfs.

8 inch

$$Q = .435 \ C \ d^{2.63} \ s^{.54}$$

$$3 = .435 \ \ 100 \ \ .67^{2.63} \ s^{.54}$$

$$.0497 = s$$

$$HL = .0497 \ \text{ft./ft.} \ \text{x} \ 1303 \ \text{ft.} = 64.8 \ \text{ft.}$$

6 inch

$$Q = .435 \ C \ d^{2.63} \ s^{.54}$$

$$3 = .435 \ \ 100 \ \ .5^{2.63} \ s^{.54}$$

$$.2068 = s$$

$$HL = .2068 \ \text{ft./ft.} \ \text{x} \ 175 \ \text{ft.} = 36.2 \ \text{ft.}$$

Adding HL's: 64.8 ft. + 36.2 ft. = 101 ft. total HL (from assumed flow)

Choose equivalent pipe: 6 inch diameter C=100.

$$s = \frac{HL}{L}$$

$$.2068 = \frac{101}{L}$$

$$488.4 \ \text{ft.} = L$$

Use L of equivalent pipe and given HL - solve for slope.

$$s = \frac{HL}{L} = \frac{50}{488.4} = .1024 \text{ ft./ft.}$$

$$Q = .435 \; C \; d^{2.63} \; s^{.54}$$

$$Q = .435 \; 100 \; .5^{2.63} \; s^{.54}$$

$$Q = 2.1 \text{ cfs}$$

CHAPTER 10

OPEN CHANNEL FLOW

Up to this point we have concentrated on water flowing in full pipes, under pressure. But what of rivers, streams, aqueducts, drainage and irrigation ditches, and sewer pipes, which are usually only partially full? Open Channel Flow exists in conduits where the water has a free surface exposed to atmospheric pressure. Water depth being constant from one point to another, there is no pressure to act as the driving force. The only natural energy this water possesses is its Velocity Head, and at normal water velocities, this is a small value ($V^2/2g$). The basic hydraulic principles apply, however. Open channels have a degree of roughness, just as water pipes under pressure do, and friction occurs as the moving water rubs against the sides of the channel. This quickly depletes the Velocity Head, and decreases the velocity. Therefore, open channel systems (aqueducts, sewers, drains) are constructed with a slope; the pipe is physically slanted downward just enough to overcome friction losses, and keep the velocity (and flow) constant. As long as the slope of the channel (ft./ft.) is exactly the same as the friction loss (ft./ft.), velocity will be maintained. Man here is making use of gravity to overcome friction, so that the Velocity Head can keep the water moving.

In a sloped channel, at steady state flow, the Hydraulic Grade Line is parallel to the water surface, and the channel bottom. It is not registering a decrease in energy. Energy loss has been negated by the slope of the channel. In this diagram, note that the depth of the water remains the same. There is no energy change - no input due to depth (Pressure Head). Pressure Head and Velocity Head are the same at Point 1 and Point 2, and the slope of the channel (the open channel equivalent of Elevation Head) is overcoming the friction ($Z_1 - Z_2 = HL_f$).

If the channel were horizontal, however, the water would decrease in velocity as it encounters friction, and "pile up" behind, making its own slope. The water produces for itself the Pressure Head needed to keep moving, as a greater depth at the upstream end, and the slope of the water surface will register its progressive loss downstream. The Hydraulic Grade Line is still drawn parallel to, and along with, the water surface,

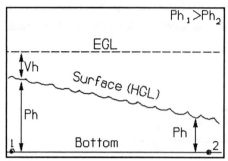

and it registers the loss of energy. The Energy Grade Line remains horizontal: flow is steady state; as cross sectional area decreases, velocity increases.

Take this channel and slant it now - downward at exactly the same slope as the water surface registers, and the surface will flatten out and become parallel with the conduit bottom again (like the first diagram). Hydraulic conditions will prove Bernoulli's Theory true.

If the channel is sloped more than is necessary to overcome friction, velocity will increase, and water depth will decrease. The extra energy created by the steep slope becomes velocity energy. Enough velocity occurs to make the friction head loss equal to the elevation drop across the length of pipe ($Z_1 - Z_2 = HL_f$). As long as the flow approaching the system remains steady, the depth will decrease substantially to accommodate the new velocity.

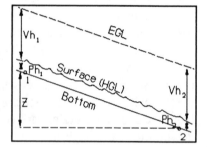

If, in an existing channel, the flow increases, the area of flow will increase, thus holding the velocity constant. Note that in each of these situations, there is a natural adjustment of the water depth, resulting in a change in the cross sectional area of flow. The Equation of Continuity (Q=AV) also always applies.

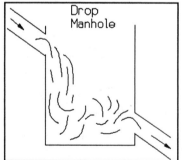

In wastewater collection systems, where the desired slope between manholes is not sufficient to maintain scouring velocity, additional slope must be provided. On the other hand, if the slope of the pipe is too steep, velocity must be reduced, and often a drop manhole is used - a method of wasting energy. Velocity suddenly drops to zero at the manhole, and for hydraulic measurements, the exit pipe is considered the upstream end of an entire new pipe system.

In natural open channels, the same phenomena are evidenced. A stream bed which slopes sharply, then levels off somewhat, will carry water at a shallow depth where the slope is steep, and velocity is high. Downstream, where the slope is gradual, the water will be deeper, and the velocity slower. Where the gradual slope was not enough to overcome friction loss, in this section the extra depth will provide enough Pressure Head at the upstream end to keep the water moving forward at a constant flow, and the depth will gradually decrease farther downstream as energy is dissipated.

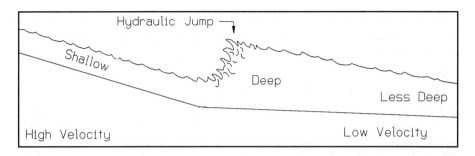

It is interesting to note that in places where waters of two different velocities meet, a short section of deeper water occurs; the water level rises at the point where the velocity changes, and a "hump" forms here, before the surface smoothes out again. There is extra turbulence at the point where the two velocities meet, and the water traps air and expands somewhat. Called Hydraulic Jump, this phenomenon occurs dramatically at dams and flumes, as described in Chapter 1, and much less dramatically where velocity change is slight. On a small scale it is visible as ripples that occur when a stone is thrown into water, or when water pours from a faucet into a full sink or bucket.

This brings up an important distinction when considering open channel flow as opposed to closed pipe flow. In a pressurized pipe, the upstream pressure determines the velocity with which the water will travel, and that velocity determines the head loss. The amount of water admitted to the system (the flow) is dependent upon this velocity and pipe diameter. In open channels, the head loss is equal to the elevation change produced by the slope of the pipe; this will determine the velocity and the depth of water in the pipe, and this velocity and depth will vary, depending upon the adequacy of the pipe slope to overcome friction losses. The total amount of flow approaching the system is admitted, and the cross sectional area of flow along with the water velocity will adjust to accommodate this flow (unless the channel isn't deep enough, and then the water will flood out over both sides).

CALCULATIONS FOR OPEN CHANNEL FLOW

In design and construction of open channels, maintenance of proper velocity is of prime importance. In order to keep the velocity constant, the channel slope must be adequate to overcome friction losses. Calculation for head loss at a given flow is necessary, and the Hazen-Williams formula is useful.

$$\text{Recalling: } Q = .435 \ C \ d^{2.63} \ s^{.54}$$

The concept of slope has not changed. The only difference is that now we are measuring, or calculating for, the physical slope of a channel (ft./ft.), equivalent to head loss.

The problem that arises is with the diameter. In a conduit which is not circular (rivers, grit chambers), or in pipes only partially full (sewers, drains), where the cross sectional area of the water is not circular, there is no diameter. Another parameter must be used to designate the size of the cross section, and the amount of it that contacts the sides of the conduit. For these non-circular areas, we use <u>Hydraulic Radius</u> (R). It is a measure of the efficiency with which the conduit can transmit water. Its value depends on pipe size, and amount of fullness. Hydraulic Radius is determined by dividing the cross sectional area of the water, by the perimeter of conduit in contact with that water.

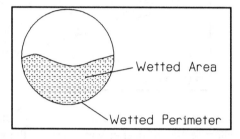

$$R = \frac{\text{wetted area}}{\text{wetted perimeter}}$$

The attempt is to measure how much of the water is in contact with the sides of the channel - or - how much of the water is not in contact with the sides, the "hydraulic" radius. The original version of the Hazen-Williams formula did incorporate Hydraulic Radius, in fact, as did similar versions developed by Chezy, Manning, and others. The formula we have been using, employing diameter, is a shortened version specialized for full pipe flow. For use in open channels, <u>Manning's Formula</u> has become most popular:

$$V = \frac{1.486}{n} R^{.66} \text{ x } s^{.5}$$

Diameter has been replaced by R, thus making the formula flexible enough to adapt to all cross sectional areas.

Hydraulic Radius for Full Pipes

Manning's formula can be used for full pipe flow also.

For a full pipe

$$R = \frac{\text{wetted area}}{\text{wetted perimeter}} = \frac{3.14 \, r^2}{3.14 \, d} = \frac{(d/2)^2}{d} = \frac{d}{4}$$

$$R = \frac{d}{4}$$

Hydraulic Radius for Different Sized Channels

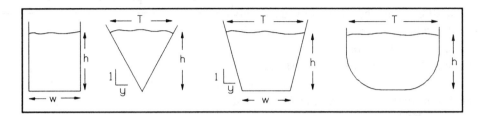

Since open channel flow includes both natural and man-made conduits, shapes can vary a great deal. Above are some typical cross sections, and below are a series of formulas to handle them in solving for R, etc.

OPEN CHANNEL SHAPES AND FORMULAS

Cross Section	Rectangle	Triangle	Trapezoid	Parabola
Area, A	wh	yh^2	$wh + yh^2$	$\dfrac{2}{3}hT$
Wetted Perimeter, P	$w + 2h$	$2h\sqrt{y^2 + 1}$	$w + 2h\sqrt{y^2 + 1}$	$T + \dfrac{8h^2}{3T}$
Hydraulic Radius, R	$\dfrac{wh}{w + 2h}$	$\dfrac{yh}{2\sqrt{y^2 + 1}}$	$\dfrac{wh + yh^2}{w + 2h\sqrt{y^2 + 1}}$	$\dfrac{2hT^2}{3T^2 + 8h^2}$
Top Width, T	w	$2yh$	$w + 2yh$	$\dfrac{3A}{2h}$

T = Top width of water surface
w = Bottom width of channel
h = Depth of water
y = Deals with the slope of the side of the channel in triangles and trapezoids. The side slopes y feet out for every one foot of depth.

These formulas eliminate the need for trigonometry, and can be referred to whenever needed.

Hydraulic Radius for Partially Full Pipes

The circular conduit is certainly the most frequently used open channel. Sewers are almost all circular - but they are usually only partially full of water. This leaves us with an odd cross section for which we must find Hydraulic Radius. Solving for cross sectional area and wetted perimeter involves more complex mathematics which can be avoided by using a <u>Hydraulic Elements Curve</u> (end of chapter). This is an indirect method of solving for wetted area, wetted perimeter,

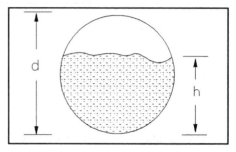

Hydraulic Radius, flow or velocity, for any size pipe, at any amount of fullness. <u>It does not give us these elements directly.</u> For instance, in looking up R for our partially full pipe, the curve yields the percent that our R is, compared to the R of a full pipe of this size. To use the curve, all we need to measure is the depth of the water in the pipe. The left hand side of the curve is Percent Depth of Flow. To obtain this, measure the depth, then divide it by the full depth (pipe diameter).

Directions for Using Hydraulic Elements Curve

-- Obtain percent depth of flow (h/d).
-- Follow across to the value you want, then down to Percent of Value for Full Pipe.
-- Solve for this value in a full pipe.
-- Multiply by the percent of value you have taken from the curve.

Inspecting this curve again, we note that:

-- When a pipe is 50% full, the values of area, wetted perimeter, and flow are half of that of a full pipe. The Hydraulic Radius and velocity, however, are the same as that of a full pipe.
-- When a pipe is between 50% and 100% full, the velocity and Hydraulic Radius are greater than they are in a full pipe.

Inspect the partially full pipes below:

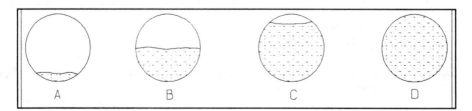

R is the relationship of wetted area to wetted perimeter. In pipe A there is a greater wetted perimeter and less wetted area. In B and D the relationship between the two is the same: half the wetted area to half the wetted perimeter (B), or all the wetted area

to all the wetted perimeter (D). In C, there is more wetted area and less wetted perimeter. So for C, the Hydraulic Radius, the amount of water that is not in contact with the sides of the pipe, is greater than its value in a full pipe. Maximum R is at about 80% fullness. It makes sense that velocity follows the characteristics of Hydraulic Radius on the curve, for it is the velocity that is decreased by friction. The Hydraulic Elements Curve is very easy to read, but we must keep in mind that it only gives us a percentage of the value for a full pipe.

Roughness

$$\text{Inspecting Manning's formula again:} \quad V = \frac{1.486}{n} R^{.66} s^{.5}$$

There is a new component - n. This is a roughness coefficient (similar to C). Materials of which open channels are constructed can be very different from those used in closed pipe flow, especially when we consider that nature has made many of them. Man-made drainage channels are often lined with concrete, asphalt, rip-rap, or grass to prevent erosion from the flowing water. Drainage through the sides and bottom is often allowed, to prevent cracking from frost heaves or hydrostatic pressure (this quantity of seepage is very minimal when compared with the quantity of flow that the conduit is carrying, however, and will not interfere with our necessity to consider a steady state flow system). Grass lined channels are used in parks and golf courses, where aesthetics are important, or where large flows are infrequent. Curb and gutter channels are designed into paved streets, and are one-sided. They must provide scouring velocities, and have enough depth so that the water surface does not interfere with traffic. Sewer pipes can be lined or unlined, and are constructed of iron, steel, concrete, brick, stones, clay and many other materials. Manning's n factors developed as a method to approximate roughness in open channels. Note that the values decrease as the channels get smoother. If equivalence is to be made to C factors, a special table at the back includes some values, though it is not very specific, and wide ranges are offered. Determination of an accurate roughness coefficient for a particular conduit is only approximate.

Velocity

Manning's formula equates to velocity, instead of flow. Perhaps this originated from the importance of providing scouring velocity in conduits designed to carry solids-laden water, and from the need to determine a slope that would provide that velocity. Solids allowed to deposit along the length of the channel will change the velocity, and eventually block the channel. If desired, it is easy to replace V with Q by using the Equation of Continuity (Q=AV, V=Q/A).

Flow

There is a most important feature of open channel flow which must be kept in mind, and which makes it quite different in practice from closed pipe flow. For calculations we always consider the flow as steady, but the actual discharge into these channels is almost never constant, and varies over the course of a day, week, and with the seasons. Most sanitary sewers are designed so that they are half full at average daily flow. Storm drainage channels are not in use at all most of the time, yet must have extremely high capacity during wet weather. This makes design quite a challenge, and operation is not often at design efficiency.

If desired, to save time and eliminate some calculations in dealing with Manning's formula, a nomograph is provided.

MANNING ROUGHNESS COEFFICIENT, n

Type of Conduit	n
Pipe:	
Cast Iron, coated	0.012 - 0.014
Cast Iron, uncoated	0.013 - 0.015
Wrought Iron, galvanized	0.015 - 0.017
Wrought Iron, black	0.012 - 0.015
Steel, riveted and spiral	0.015 - 0.017
Corrugated	0.021 - 0.026
Wood Stave	0.012 - 0.013
Cement Surface	0.010 - 0.013
Concrete	0.012 - 0.017
Vitrified	0.013 - 0.015
Clay, drainage tile	0.012 - 0.014
Lined Channels:	
Metal, smooth semicircular	0.011 - 0.015
Metal, corrugated	0.023 - 0.025
Wood, planed	0.010 - 0.015
Wood, unplaned	0.011 - 0.015
Cement Lined	0.010 - 0.013
Concrete	0.014 - 0.016
Cement Rubble	0.017 - 0.030
Grass	----- - 0.200
Unlined Channels:	
Earth: straight and uniform	0.017 - 0.025
dredged	0.025 - 0.033
winding	0.023 - 0.030
stony	0.025 - 0.040
Rock: smooth and uniform	0.025 - 0.035
jagged and irregular	0.035 - 0.045

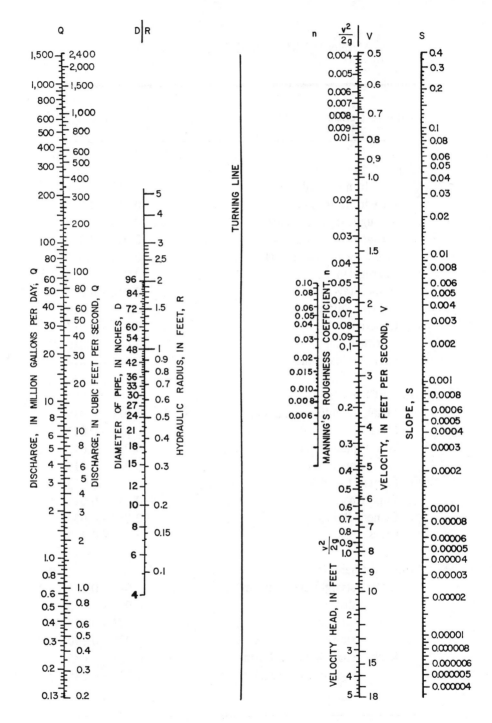

Manning's Nomograph

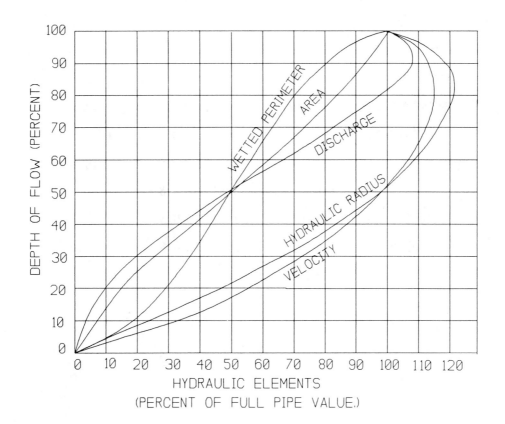

Conduit Material	Manning n (ft.)	Hazen-Williams C
Closed conduits		
Asbestos-cement pipe	0.011 - 0.015	100 - 140
Cast iron pipe cement-lined	0.011 - 0.015	100 - 140
Concrete pipe	0.011 - 0.015	100 - 140
Corrugated-metal pipe, asphalt lined	0.011 - 0.015	100 - 140
Plastic pipe (smooth)	0.011 - 0.015	100 - 140
Vitrified clay pipe	0.011 - 0.015	100 - 140

PROBLEMS

1. Water flows at a depth of 6.25 ft. in a rectangular canal 8 ft. wide. The average water velocity is 14.9 ft./sec. What is the slope on which the canal is laid if n=.01?

2. Determine the drop in elevation from beginning to end of a rectangular wooden flume 1000 ft. long, which carries 600 cfs at a 7 ft. depth. The flume measures 15 ft. across.

3. A rectangular channel whose width is three times its depth is lined with smooth concrete. When carrying a flow of 900 cfs, the average velocity is 3 ft./sec. What is the incline of this channel?

4. A 12 inch diameter lined cast iron sewer flowing full is laid on a grade of .004 ft./ft. Calculate the flow:
    A. using Manning's formula
    B. using Hazen-Williams formula

5. A reservoir feeds an open rectangular channel 15 ft. wide (n=.015). The depth of water in the reservoir is 6.22 ft. above the channel bottom. The channel is 800 ft. long, and drops .72 ft. in this length. The depth behind a weir at the discharge end of the channel is 4.12 ft. What is the flow in the channel?

6. A 48 inch diameter circular steel drain is laid on a slope of .003, and is flowing full. What is the flow?

7. The cross sectional measurements of a flume are 10 inches by 10 inches. The water is 8 inches deep, and moving at a velocity of 2 ft./sec. How many gallons of water will the flume deliver in four hours?

8. Water flows 3 ft. deep in a rectangular channel 20 ft. wide (n = .013, s = .0144). How deep would the same quantity flow on a slope of .00144?

9. Which of these canal structures will carry the greater flow if both are laid on the same slope?

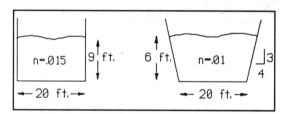

10. What is:
    A. the Hydraulic Radius
    B. the cross sectional area for a flow of 6 inches depth in a 10 inch diameter circular pipe?

11. A 10 inch diameter pipe is flowing 1/3 full. What is the velocity of the water if the volume of flow is 135 gpm?

12. A 6 ft. deep flow is carried by a triangular channel with a side slope of 4 ft./ft. depth. If n=.025 and s=.006, at what velocity is the water flowing?

13. A 10 inch diameter sewer pipe with a 3 inch water depth is laid on a grade of two feet per hundred feet of length (n=.025). What is the flow?

14. A 48 inch diameter concrete pipe carries a flow of 43 cfs on a slope of 1.8 ft./1000 ft. How deep is the water in the pipe?

15. A. Determine the required diameter for a proposed sanitary sewer which must carry domestic sewage at 2.5 cfs. This maximum flow is estimated to be 250% of average daily flow. Pipe to be used is PVC (n=.009).
    B. What will be the depth of flow at average and maximum flow conditions?
    C. What is the water velocity at both flows?

16. An obstruction in a grass lined channel 6 ft. wide causes a depth buildup of 2 ft. behind it. The channel bottom is horizontal and the flow is 10 MGD. How deep is the water 1000 ft. upstream?

17. A clay drainpipe flowing full slopes downward 3 ft. over a length of 1500 ft. If the water velocity in the pipe is 5 ft./sec., what is the diameter of the pipe?

18. Water from a filter leak collects on the floor beneath, then flows to a 6 ft. diameter sump from which it is carried by 150 ft. of 12 inch diameter pipe to a drainage area (n = .012 for the pipe). Free flow exists at the pipe outlet. The floor elevation is 112.0, invert of the pipe at entrance is 106.0, and outlet is 103.0. Bottom of sump is 104.0. The water level in the sump stabilized at one foot below floor level. How many gpm are being lost from the filter leak?

19. A proposed sanitary sewer must carry a flow of 125 cfs. Concrete pipe will be used. The street has a general slope of .02%, which will be followed. The minimum allowable velocity will be 1.5 ft./sec.
    A. Determine the required pipe diameter
    B. Determine the depth of water in the pipe

20. A 100 MGD water supply will be carried by gravity flow aqueduct from a mountain reservoir 150 miles to the city for treatment. Design the transport system.

SOLUTIONS

1. **Answer**   .0031 ft./ft.

Draw Diagram. Solve for R, then slope.

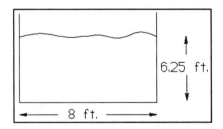

$$R = \frac{WA}{WP} = \frac{6.25 \times 8}{6.25 + 8 + 6.25} = 2.44 \text{ ft.}$$

$$V = \frac{1.486}{n} \ R^{.66} \ s^{.5}$$

$$14.9 = \frac{1.486}{.01} \ 2.44^{.66} \ s^{.5}$$

.0031 ft./ft. = s

2. **Answer**   .4 ft.

Draw Diagram. Solve for R, then Vel.,
then slope, then drop.

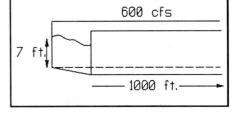

$$R = \frac{WA}{WP} = \frac{7 \times 15}{7 + 15 + 7} = 3.62 \text{ ft.}$$

Q = A   V

$600 = 15 \times 7 \times V$

5.71 ft./sec. = V

$$V = \frac{1.486}{n} \; R^{.66} \; s^{.5}$$

$$5.71 = \frac{1.486}{.012} \; 3.62^{.66} \; s^{.5}$$

.0004 ft./ft. = s

.0004 ft./ft. x 1000 ft. = .4 ft. of incline

3. **Answer**   .00009 ft./ft.

Draw Diagram. Solve for area, then
width and depth; solve for R, then
slope.

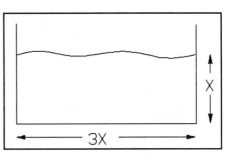

Q = A   V

$900 = A \times 3$

300 sq.ft. = A

Area = width x depth

$300 = 3X \times X$

10 ft. = X

$$R = \frac{WA}{WP} = \frac{10 \times 30}{10 + 30 + 10} = 6 \text{ ft.}$$

$$V = \frac{1.486}{n} \; R^{.66} \; s^{.5}$$

$$3 = \frac{1.486}{.015} \; 6^{.66} \; s^{.5}$$

.00009 ft./ft. = s

4. **Answer**   A. 2.3 cfs
                 B. 2.6 cfs

A. <u>Manning</u>: Solve for R, then Vel., then flow.

$$R = \frac{d}{4} = \frac{1}{4} = .25 \text{ ft.}$$

$$V = \frac{1.486}{n} \quad R^{.66} \quad s^{.5}$$

$$V = \frac{1.486}{.013} \quad .25^{.66} \quad .004^{.5}$$

$$V = 2.89 \text{ ft./sec.}$$

$$Q = A \quad V$$

$$Q = .785 \times 1 \times 2.89$$

$$Q = 2.3 \text{ cfs}$$

B. <u>Hazen-Williams</u>: Refer to equivalence chart for roughness coefficients for C factor; solve for flow.

$$C = 120$$

$$Q = .435 \quad C \quad d^{2.63} \quad s^{.54}$$

$$Q = .435 \quad 120 \quad 1 \quad .004^{.5}$$

$$Q = 2.6 \text{ cfs}$$

---

5. **Answer**   692 cfs

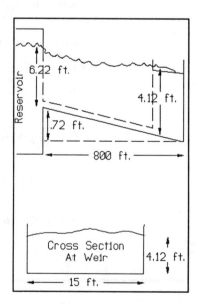

Draw Diagram. Assume unrestricted flow from the reservoir. HL is equivalent to total slope of channel plus extra depth at reservoir. Solve for slope, then R, then velocity, then flow.

HL = .72 ft. drop of channel + 6.22 ft. - 4.12 ft. depth of reservoir = 2.82 ft.

$$s = \frac{HL}{L} = \frac{2.82}{800} = .0035 \text{ ft./ft.}$$

$$R = \frac{WA}{WP} = \frac{4.12 \times 15}{4.12 + 15 + 4.12} = 2.66 \text{ ft.}$$

$$V = \frac{1.486}{n} \ R^{.66} \ s^{.5}$$

$$V = \frac{1.486}{.015} \ 2.66^{.66} \ .0035^{.5}$$

$$V = 11.2 \text{ ft./sec.}$$

$$Q = A \ V$$

$$Q = 4.12 \times 15 \times 11.2$$

$$Q = 692 \text{ cfs}$$

6. **Answer** 63.9 cfs

Draw Diagram. Solve for R, then Vel., then flow.

$$R = \frac{d}{4} = \frac{4}{4} = 1 \text{ ft.}$$

$$V = \frac{1.486}{n} \ R^{.66} \ s^{.5}$$

$$V = \frac{1.486}{.016} \ 1 \ .003^{.5}$$

$$V = 5.09 \text{ ft./sec.}$$

$$Q = A \ V$$

$$Q = .785 \times 4^2 \times 5.09$$

$$Q = 63.9 \text{ cfs}$$

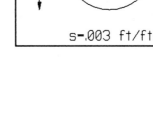

48 in.   Full

s-.003 ft/ft

7. **Answer** 118,483 gal./4 hrs.

Draw Diagram.

$$Q = A \ V$$

$$Q = .83 \times .67 \times 2$$

$$Q = 1.1 \text{ cfs}$$

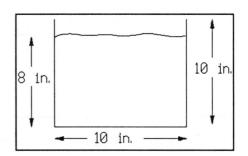

8 in.

10 in.

10 in.

1.1 cuft/sec. x 60 sec./min x 60 min./hr. x
7.48 gal./cuft x 4 hr. = 118,483 gal./4 hr.

8. **Answer**   6.5 ft.

Draw Diagram. Solve for V, then Q at
s=.0144.   Create a comparison chart
at different depths; solve for Vel.
through both Mannings and Q=AV. At
a given depth, if the Vel. comes out the
same through both formulas, then that
must be the right depth.

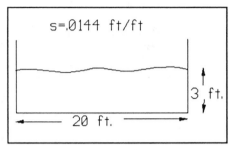

$s = .0144$

$$R = \frac{WA}{WP} = \frac{3 \times 20}{3 + 20 + 3} = 2.31 \text{ ft.}$$

This is R at a depth of 3 ft. - which the slope
of .0144 ft./ft. creates.

$$V = \frac{1.486}{n} \; R^{.66} \; s^{.5}$$

$$V = \frac{1.486}{.013} \; 2.31^{.66} \; .0144^{.5}$$

$$V = 23.8 \text{ ft./sec.}$$

$$Q = A \; V$$

$$Q = 20 \times 3 \times 23.8$$

$$Q = 1428 \text{ cfs}$$

$s = .00144$

Depth Trials:
Assume a depth: using it, solve for Velocity using Mannings (s=.00144),
then using Q=AV. When velocities match, then depth is correct depth.

Depth (ft.)	Velocity (ft./sec.) (Mannings)	Velocity (ft./sec.) (Q=AV)
10	12.5	7.1
15	14.1	4.8
6.5	10.7	10.9

9. **Answer**   Trapezoid carries greater flow.

Both are laid on the same slope - so assume any slope. Solve for R, then Velocity,
then flow in each.

Assume slope = 1

Rectangle:

$$R = \frac{WA}{WP} = \frac{9 \times 20}{9 + 20 + 9} = 4.74 \text{ ft.}$$

$$V = \frac{1.486}{n} \quad R^{.66} \quad s^{.5}$$

$$V = \frac{1.486}{.015} \quad 4.74^{.66} \quad 1$$

$$V = 277 \text{ ft./sec.}$$

$$Q = A \quad V$$

$$Q = 9 \times 20 \times 277$$

$$Q = 49860 \text{ cfs}$$

Trapezoid:

$$R = \frac{wh + yh^2}{w + 2h\sqrt{y^2 + 1}} = \frac{(20 \times 6) + (1.33 \times 6^2)}{20 + (2 \times 6)\sqrt{1.33^2 + 1}} = 3.15 \text{ ft.}$$

$$V = \frac{1.486}{n} \quad R^{.66} \quad s^{.5}$$

$$V = \frac{1.486}{.01} \quad 3.15^{.66} \quad 1$$

$$V = 317 \text{ ft./sec.}$$

$$Q = A \quad V$$

$$Q = wh + yh^2 \quad V$$

$$Q = (20 \times 6) + (1.33 \times 6^2) \times 317$$

$$Q = 53218 \text{ cfs}$$

---

10. **Answer**   A. .2283 ft.
                 B. .341 sq.ft.

Draw Diagram.

A. Obtain % fullness; refer to Hydraulic Elements Curve for % R of full pipe; solve for R of full pipe; multiply by % value for our pipe.

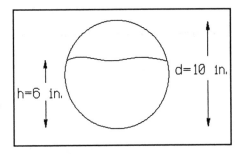

% fullness = 6/10 = 60%

From curve, R for a 60% full pipe is 110% of R of a full pipe.

$$R \text{ (full pipe)} = \frac{d}{4} = \frac{.83}{4} = .2075 \text{ ft.}$$

.2075 ft. (R-full) x 110% (our % of full value) = .2283 ft. (R - our pipe)

B. Do the same thing for Area.

Area of 60% full pipe = 63% area of full pipe.

$A \text{ (full pipe)} = .785 \quad d^2$

$A = .785 \times .83^2$

$A = .54 \text{ sq.ft.}$

.54 sq.ft. (A - full) x .63 (our % of full value) = .341 sq.ft. (A - our pipe)

11. **Answer**   2 ft./sec.

Change gpm to cfs; refer to Hyd.El.Curve for % A; obtain value for A; solve for Vel.

$$\frac{135 \text{ gpm}}{60 \times 7.48} = .3 \text{ cfs}$$

33% full pipe: A = 28% of A (full pipe) - from curve.

$A \text{ (full)} = .785 \quad d^2$

$A = .785 \times .83^2$

$A = .54 \text{ sq.ft.}$

.54 sq.ft. (A - full) x .28 (our % of full value) = .15 sq.ft. (A - our pipe)

$Q = A \quad V$

$.3 = .15 \times V$

$2 \text{ ft./sec.} = V$

12. **Answer**   9.3 ft./sec.

Draw Diagram. Solve for R, then V.

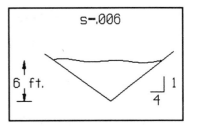

$$R = \frac{yh}{2\sqrt{y^2 + 1}}$$

$$R = \frac{4 \times 6}{2\sqrt{4^2 + 1}}$$

R = 2.91 ft.

$$V = \frac{1.486}{n} \; R^{.66} \; s^{.5}$$

$$V = \frac{1.486}{.025} \; 2.91^{.66} \; .006^{.5}$$

V = 9.3 ft./sec.

13. **Answer**   .31 cfs

Draw Diagram. Solve for slope; using Hyd.E1.Curve, solve for R, then Velocity. Using curve again, solve for Area, then flow.

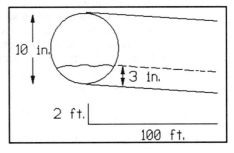

$$\% \text{ fullness} = \frac{3}{10} = 30\% \text{ full}$$

With curve, 30% full pipe:
R = 67% of R for full pipe.

$$R \text{ (full pipe)} = \frac{d}{4} = \frac{.83}{4} = .2075 \text{ ft.}$$

.2075 ft. (R - full) x .67 (% full our pipe) = .139 ft. (R - our pipe)

$$V = \frac{1.486}{n} \; R^{.66} \; s^{.5}$$

$$V = \frac{1.486}{.025} \; .139^{.66} \; .02^{.5}$$

V = 2.29 ft./sec.

With curve, 30% full pipe: A = 25% of A for full pipe.

$$A \text{ (full)} = .785 \quad d^2$$

$$A = .785 \text{ x } .83^2$$

$$A = .54 \text{ sq.ft.}$$

.54 sq.ft. (A - full) x .25 (% full our pipe) = .135 sq.ft. (A - our pipe)

$$Q = A \quad V$$

$$Q = .135 \text{ x } 2.29$$

$$Q = .31 \text{ cfs}$$

---

14. **Answer**   32 inches deep

Draw Diagram. Solve for slope. There is too little data to solve for depth directly either by Q=AV or by Mannings formula. If we <u>had</u> the correct depth of flow, we could obtain velocity by either formula, and it should be the same velocity by either. Assume different depths; create a chart. Compare velocities which each formula yields. When they are the same, then that is the correct depth.

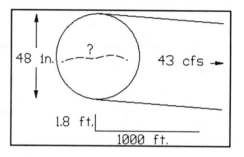

Depth	% Fullness	R	V (Manning)	V (Q=AV)	A
24 in.	50%	1	4.5	6.9	6.28
36 in.	75%	1.2	5.1	4.3	10
48 in.	100%	1	4.5	3.4	12.6
32 in.	65%	1.1	4.9	4.9	8.2

It is necessary to work with the Hyd.Eq.Curve through % fullness to a true value for R & A for our assumed depth.

After trying 50%, 75%, 100% fullness, and comparing velocities, 65% was attempted.

---

15. **Answer**   A. 14 inch diam.
      B. Max. Q: 10.4 inches deep; Av. Q: 6 inches deep
      C. Max. Q: Vel. = 3 ft./sec.; Av. Q: Vel. = 2.3 ft./sec.

A. Sanitary sewer must have scouring velocity. Start with Vel. = 3 ft./sec. Since max. flow is 2.5 cfs, let pipe be full at this flow. Solve for diameter.

$$Q = A \quad V$$

$$2.5 = .785 \quad d^2 \quad 3$$

$$1.03 \text{ ft.} = d$$

Choose 14 inch pipe. (1.167 ft.)

B. We will slope our pipe enough to keep the velocity at 3 ft./sec. The area of flow is based on the full pipe diameter calculated above; solve for this area of flow. Get its % of the full area of our pipe; refer to Hyd. Elements Curve for % depth of flow.

$$A = .785 \quad d^2$$

$$A = .785 \text{ x } 1.03^2$$

$$A = .833 \text{ sq.ft.}$$

$$A = .785 \quad d^2$$

$$A = .785 \text{ x } 1.167^2$$

$$A = 1.07 \text{ sq.ft.}$$

$$\frac{.833 \text{ sq.ft.}}{1.07 \text{ sq.ft.}} = .78 = 78\% \text{ of area of pipe is submerged}$$

From Hyd. Elements Curve 78% of area = 74% depth of flow. (in our pipe)

.74 x 1.167 ft. diam. = .864 ft. depth = 10.4 inches deep

Average daily flow = 1cfs. This is 40% of max. flow (2.5 cfs).

Max. flow fills 74% depth of pipe.

From Hyd. Elements Curve, Max. flow, at this depth, is 90% of the flow of this pipe full.

Therefore, our average flow = 40% x 90% of flow of full pipe = 36% of flow of full pipe.

Again, from curve, at 36% of flow of full pipe, (our average flow), the pipe holds 43% depth of flow.

.43 x 1.167 ft. diam. .5 ft. depth = 6 inches deep

C. Max flow: 78% of area of full pipe area. Solve for velocity.

$$Q = A \quad V$$

$$2.5 = .78 \text{ x } .785 \text{ x } 1.167^2 \quad V$$

$$3 \text{ ft./sec.} = V$$

Note: Area of flow is the same as in original pipe diameter calculation; therefore, velocity is the same.

Average flow: at 43% depth of flow in pipe, area (from curve) is 41% of area of full pipe. Solve for velocity.

Q = A   V

$1 = .41 \times .785 \times 1.167^2$ V

2.3 ft./sec. = V

---

16.   **Answer**   26 ft.

Draw Diagram. Since channel bottom is horizontal, water surface will create a slope which is equivalent to the HL. Depth upstream = HL/1000 ft. plus the 2 ft. obstruction.

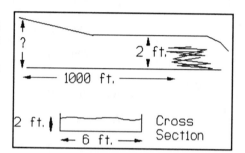

$$R = \frac{WA}{WP} = \frac{6 \times 2}{2 + 6 + 2} = 1.2 \text{ ft.}$$

10 MGD = 15.5 cfs

Q = A   V

$15.5 = 2 \times 6$ V

1.3 ft./sec. = V

$$V = \frac{1.486}{n} \quad R^{.66} \quad s^{.5}$$

$$1.3 = \frac{1.486}{.2} \quad 1.2^{.66} \quad s^{.5}$$

.024 ft./ft. = s

HL = .024 ft./ft. x 1000 ft. = 24 ft.

24 ft. HL + 2 ft. for obstruction = 26 ft.

---

17.   **Answer**   4 ft.

Solve for slope; solve for R, then diameter.

$$s = \frac{HL}{L} = \frac{3}{1500} = .002 \text{ ft./ft.}$$

$$V = \frac{1.486}{n} \, R^{.66} \, s^{.5}$$

$$5 = \frac{1.486}{.013} \, R^{.66} \, .002^{.5}$$

1 ft. = R

$$R = \frac{d}{4}$$

$$1 = \frac{d}{4}$$

4 ft. = d

---

18. **Answer**   4039 gpm

Draw Diagram. HL = 3 ft. of slope plus 5 ft. of pressure in sump = 8 ft. HL. Solve for slope, then R, then velocity, then flow; change to gpm.

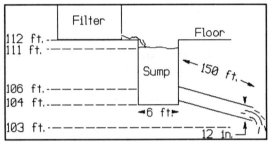

$$S = \frac{HL}{L} = \frac{8}{150} = .0533 \text{ ft./ft.}$$

$$R = \frac{d}{4} = \frac{1}{4} = .25 \text{ ft.}$$

$$V = \frac{1.486}{n} \, R^{.66} \, s^{.5}$$

$$V = \frac{1.486}{.012} \, .25^{.66} \, .0533^{.5}$$

$$V = 11.5 \text{ ft./sec.}$$

$$Q = A \quad V$$

$$Q = .785 \times 1 \times 11.5$$

$$Q = 9 \text{ cfs}$$

9 cfs x 60 sec./min. x 7.48 gal./cu.ft. = 4039 gpm

19. **Answer**   A. 12 ft. diameter
                 B. 8.4 ft. depth

  A. Solve for area of flow; choose diameter.

$$Q = A \quad V$$

$$125 = A \times 1.5$$

83.3 sq.ft. $= A$ This is area of <u>flow</u>.

$$A = .785 \quad d^2$$

$$83.3 = .785 \quad d^2$$

10.3 ft. $= d$ This is the minimum pipe size that would hold this flow.

    Therefore choose 12 ft. diameter pipe.

  B. Solve for Area of chosen pipe; solve for % of area that our flow uses in this pipe; consult Hyd. El.Curve for % fullness; solve for depth.

$$A = .785 \quad d^2$$

$$A = .785 \times 12^2$$

$$A = 113 \text{ sq.ft.}$$

$\dfrac{83.3 \text{ sq.ft.}}{113 \text{ sq.ft.}} = .74 = 74\%$ our flow is 74% of full pipe area

From curve, our pipe $= 70\%$ full.

12 ft. diam. pipe x .7 $= 8.4$ ft. depth of flow

Check velocity with Manning's formula: 3.2 ft./sec.

---

20. **Answer**   Variable

Assumptions: need to have these
Pipe material choice: cement lined steel pipe (n=.013)
100 MGD $= 155$ cfs $=$ pipe flowing full
Wish to have scouring velocity: Vel. $= 3$ ft./sec.
Solve for diameter:

$$Q = A \quad V$$

$$155 = .785 \times d^2 \times 3$$

$$8.113 \text{ ft.} = d$$

Choose 8.5 ft. diameter pipe

Calculate slope for pipe construction, using our chosen diameter pipe; obtain slope that will produce 3 ft./sec. velocity; even though this is no longer a full pipe, the area of flow is the same, therefore the velocity remains the same. However, R is not that of a full pipe; compare areas to obtain depth of flow. Get % R compared to that of a full pipe. Obtain R for our pipe; then solve for slope.

Area of flow $= .785 \times 8.113^2 = 51.67$ sq.ft.

Area of our pipe $= .785 \times 8.5^2 = 56.72$ sq.ft.

$\dfrac{51.67}{56.72} = .91 = 91\%$ of full pipe area

Referring to Hyd. Elements Curve, with 91% of full pipe area, pipe is 86% full.

From curve, a pipe 86% full has an R that is 122% of full pipe R.

Full pipe R $= \dfrac{d}{4} = \dfrac{8.5}{4} = 2.125$ ft.

Our pipe R $= 1.22 \times 2.125$ ft. $= 2.6$ ft.

Vel. $= \dfrac{1.486}{.013}$ $R^{.66}$ $s^{.5}$

$3 = 114.3$ $2.6^{.66}$ $s^{.5}$

$.0002$ ft./ft. $= s$

$.0002$ ft./ft. x 5280 ft./mile x 150 miles $= 158.4$ ft. drop in elevation needed for pipeline

CHAPTER 11

FLOW MEASUREMENT I
FLOW RATE METERS

Why measure flow? The most vital activities in the operation of water and wastewater treatment plants are dependent on a knowledge of how much water is being processed.

The municipal water works must be able to measure water made and water used in order to calculate percent lost water. Knowledge of flow is needed to size storage tanks, provide firefighting capacity, to guard against backflow, to predict future usage, and to manage water conservation programs.

Wastewater treatment plants need a flow measurement to determine solids and BOD loading to the plant, and to the receiving stream for NPDES compliance. Control of recycle rates, calculation of F/M ratio, and prediction of physical and biological unit performance is based on flow. Knowledge of incoming water quantity is vital to anticipation of plant reaction to industrial input, shock loads, and stormwater. The decision to use stormwater retention basins to prevent flooding and backup is based on a flow reading. The smooth operation of an Industrial Pretreatment Program depends on knowing the quantity of industrial input and how it compares to the quantity of the municipal stream.

Both water and wastewater utilities need flow measurements to calculate chemical feed rates, hydraulic loading to treatment units, tank detention times. It is necessary for scheduling of system maintenance, for calculation of process expenses (power, chemicals, labor), for sizing of pipes, pumps and meters. Flow records are needed to plan treatment plant expansions, to be prepared for legal proceedings, and of course, for customer billing.

Chapters 11 and 12 deal with the instruments that measure flow - the flowmeters.

Flow measurement can be based on flow rate, or flow amount:

Flow Rate (gpm, MGD, cfs): Utilities need this type of meter inside the treatment plant, in wastewater collection, and in potable water distribution, to determine process variables. In common use are pressure differential meters, magnetic meters, and ultrasonic meters. Some are designed for metering flow in closed pipe systems; some are meant for open channel flow.

Flow Amount (gal., cu.ft.): The totalizer - it sums up the gallons or cubic feet which pass through the meter. Private, commercial and industrial customer service meters operate this way, and total amount of flow is the basis for billing. In wastewater treatment plants, automatic composite sampling units are usually flow proportioned to grab a sample every so many gallons. These totalizer meters may be turbine, propeller, compound or positive displacement.

We will consider flow rate meters in this chapter, and totalizers in Chapter 12.

PRESSURE DIFFERENTIAL METERS - CLOSED PIPE FLOW

Orifice Meters

Orifice meters have been in use since ancient Roman times in water pipes. The principle

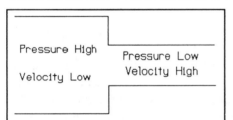

of operation is that a pressure differential is created across the unit as a steady flow of water passes through a constricted section. From Chapter 5 we recall that as water flows from an area of large cross section to an area of small cross section, the velocity increases (Q=AV). Since this results in an increase in Velocity Head, and total energy must be maintained, some form of energy must decrease by the same amount. It is the Pressure Head that drops abruptly at this point. We apply this in practice to measure flow. The sensing unit of the orifice meter is a flat metal plate, usually stainless steel, installed between flanges in a pipeline, with a precisely sized small diameter hole at the center. Pressure gages are installed one pipe diameter upstream, (ahead of turbulence at the plate), and one half pipe diameter downstream (where the jet is narrowest and pressure is minimal) to measure the pressure

differential. The formula which calculates flow for an orifice meter is nothing more than a rearrangement of Bernoulli's Theory.

Bernouilli's Formula
$$Ph_1 + Vh_1 + Z_1 = Ph_2 + Vh_2 + Z_2 + HL_f$$

Eliminate Z_1 & Z_2, and consider Vh_1 as negligible
$$Ph_1 = Ph_2 + Vh_2 + HL_f$$

Convert Vh_2 to $V^2/2g$
$$Ph_1 - Ph_2 = \frac{V^2}{2g} + HL_f$$

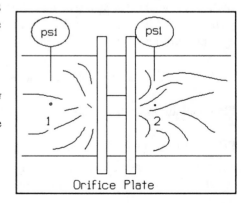

Orifice Plate

Equate to velocity
$$\sqrt{(Ph_1 - Ph_2) \times 2g} = V + HL_f$$

Equate to flow
$$A\sqrt{(Ph_1 - Ph_2) \times 2g} = Q + HL_f$$

Now the only component left out is the head loss. For ease of calculation for meters of this type, friction loss is incorporated as a Coefficient of Discharge (C_d), which takes in the loss of head encountered from passing through the meter. Its value depends upon the type of orifice chosen. Diagrammed below are several types, with their coefficients of discharge:

Sharp-Edged Orifice $C_d = .6$

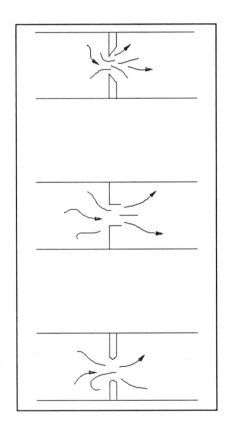

Short Tube $C_d = .8$

Round-Edged Orifice $C_d = .9$

Square shaped Orifice $C_d = .6$

In short, C_d represents a percent reduction in flow caused by turbulence at the constriction.

Including the Coefficient of Discharge, the formula becomes:

$$Q = C_d\, A\sqrt{(Ph_1 - Ph_2) \times 2g}$$

This is neat and easy to use, but keep in mind that the solution can also be arrived at by using Bernoulli's Formula, and the Equivalent Pipe for Fittings chart to calculate the head loss.

Note that the pressure drop resulting from the velocity change at the constriction is recoverable as the water continues on the discharge side of the orifice, but the head lost from turbulence created in getting through the constricted section is not.

Other useful applications of this flow formula occur wherever there is an opening or hole in a tank or pipeline. These are orifices, and flow can be calculated:

The diameter of the orifice is always used, and the formula applies even in cases where water flows freely from the structure to the atmosphere. In this situation, $Ph_2 = 0$. Ph_1 in the tank or pipe, is the pressure at that point.

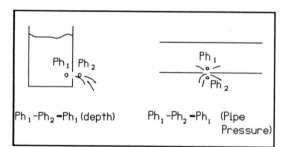

Orifice meters are inexpensive and easy to install. They require little space, and are widely used in industry for water, steam and gas flow measurements. However, there are some important disadvantages:

Orifice meters are not extremely accurate. The device causes a great amount of turbulence on discharging from the orifice which results in large head loss to the whole system. Note that the presence of the Coefficient of Discharge in the formula could decrease the flow by 40%. In addition, the greater the difference between the two diameters, the greater will be this turbulence, and resultant head loss. This is not accounted for in the formula. Small initial cost must be weighed against cost of pumping, and if high pressure and specific flow quantity are important to the system, this may not be the metering device of choice.

Corrosion and solids in the water can shortly wear a sharp edged orifice into a round edged orifice, substantially changing the C_d value (C_d goes up, but by how much?).

Orifice meters are not recommended for permanent installation to measure wastewater flows; solids in the water easily catch on the orifice, throwing off accuracy.

For installation, it is necessary to have ten diameters of straight pipe ahead of the meter to create a smooth flow pattern, and five diameters of straight pipe on the discharge side. In some installations where a meter is desired, this is not possible.

Derivation

Bernoulli's Formula:

$$Ph_1 + Vh_1 + Z_1 = Ph_2 + Vh_2 + Z_2 + HL_f$$

Eliminate Z_1 & Z_2

$$Ph_1 + Vh_1 = Ph_2 + Vh_2 + HL_f$$

$$Ph_1 - Ph_2 = Vh_2 - Vh_1 + HL_f$$

Convert to $V^2/2g$

$$Ph_1 - Ph_2 = \frac{V_2^2 - V_1^2}{2g} + HL_f$$

From $AV = AV$, we can derive

$$V_1 = \frac{d_2}{d_1} \times V_2 \quad \text{and} \quad V_1^2 = \frac{d_2^4}{2d_1} \times V_2^2$$

Substituting into formula

$$Ph_1 - Ph_2 = \frac{V_2^2 - \left(\frac{d_2}{d_1}\right)^4 \times V_2^2}{2g} + HL_f$$

Solving for V_2 yields

$$Q = C_d \ A \ \sqrt{\frac{(Ph_1 - Ph_2) \times 2g}{1 - \left(\frac{d_2}{d_1}\right)^4}}$$

Convert to Q, and add C_d

$$Q = C_d \ A \ \sqrt{\frac{(Ph_1 - Ph_2) \times 2g}{1 - \left(\frac{d_2}{d_1}\right)^4}}$$

D_1 = upstream diameter

D_2 = throat diameter

$$Q = C_d \ A \ \sqrt{\frac{(Ph_1 - Ph_2) \times 2g}{1 - \left(\frac{d_2}{d_1}\right)^4}}$$

Venturi Meters

Venturi meters, like orifice meters, measure flow in closed conduits by recording a pressure differential across a constriction in the pipe. The meter consists of a constricted tube, complete with pressure gages or manometer, and is purchased as a unit. It has a converging section, a throat (where fluid velocity increases), and a longer diverging section. The gages are placed just upstream from the convergence, and

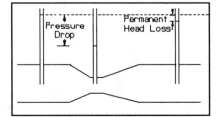

at the throat. The pattern of flow is smooth through the unit, and there is very little pressure loss due to turbulence. The angle of divergence is about 7%, which does require the tube to have substantial length. Like the orifice, the flow formula for venturi meters is derived from Bernoulli, and it was he, in the 1700's, who first described its operation. Calculation would be the same as for the orifice, except for one important difference. Since the diameters of entrance pipe and throat are not very different, the velocity heads at these two points of measurement are also not too far apart in value. Therefore, Vh_1 cannot be ignored.

The area used remains the area of the throat.

NOTE: If the throat diameter is ½ the upstream diameter, or less, the inclusion of Vh_1 is insignificant, and a good approximate flow may be obtained using the orifice formula.

The venturi meter is the most accurate and the most popular closed conduit differential pressure meter. It can be used over a wide range of flows, and has a low permanent head loss because of the smooth configuration. It is applicable to the measurement of wastewaters and sludges, for there are no protruding parts to catch solids or wear rapidly. Maintenance is a matter of flushing, and cleaning solids out of the pressure gage tubes. Calibration is done by testing at a known flow rate.

Disadvantages of the venturi are its long length, and high capital and installation costs.

Flow Rate Controller

In water filtration facilities which operate constant rate filters, the rate of flow through the filters is regulated by a valve/venturi mechanism, the Flow Rate Controller. It is installed in the filter effluent pipe, and responds to flow changes through the filter sand. As the pores in the filter media fill up with suspended matter, head loss in the bed increases, and less water flows through. At the beginning of

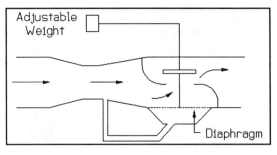

the filter run, the controller valve is almost shut. Water passing through the filter has little resistance, and flow would be high, except that passing through this almost-closed valve it encounters enough head loss to drop the rate to a predetermined set point. As the filter bed clogs, passing less and less flow through, the controller valve responds to the dropping pressure from the venturi, and gradually opens, thus keeping the flow constant.

Flow Rate Controller: Operating Mechanism

Adjustment weight on calibrated lever at top is preset to desired flow, and aids valve opening (pulls up). Pressure from flow passing through body of venturi has flexible diaphragm in depressed position at beginning of run, and valve is almost closed. Lesser pressure underneath diaphragm from venturi throat contributes to positioning. A pressure differential is created above and below diaphragm, which determines its exact position. As pressure decreases during the run, this differential decreases, and diaphragm is allowed to move up, slowly opening the valve, and maintaining the flow at a constant rate.

Modern flow rate controllers are usually computer controlled and programmable. The advantage of automatic operation is that the set point for valve movement can be adjusted from the control center, movement of the valve is more precisely controllable, and the controller can be set to sort out and respond to a number of other inputs (turbidity, temperature, initiation of backwash, high water alarm in clearwell). In the past, a series of relays was set up to respond to multiple inputs, but fluctuating response from the many inputs (searching) would result. With the Programmable Controller, a clear decision is made, and acted upon.

Differential Manometers

Manometers can be installed in venturi meters or orifices instead of the pair of pressure gages. A manometer is a piezometer tube curved in a U shape, and filled with a liquid of known Specific Gravity. Pressure on one end of the tube causes the liquid to move, by an amount equivalent to the pressure.

Simple Manometers are connected to the pressure source at one end, and the internal liquid is free to ride up the other end, as the pressure moves it. These simply read pressure. Manometers installed on anaerobic digester gas lines are of this type. They are also often seen installed on liquid chemical storage tanks, and are used as a means to determine liquid level. The internal fluid may be brightly colored for visibility. The barometer, a manometer which responds to air pressure, is of this type also. It is in contact with the atmosphere at one end (the pressure source), and the free end is enclosed in vacuum. The unit is placed up against a calibration scale so that the liquid movement can be measured.

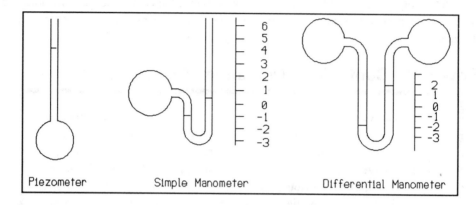

Differential Manometers are connected at both ends, to different sources of pressure. Installed into a venturi meter, the device would provide not just a pressure reading, but a pressure differential between the two sources. The center liquid can ride back and forth, depending on which side has the greater pressure. By attaching the differential manometer to both high and low pressure sides of the venturi constriction, the difference in pressure can be measured, and then converted to flow rate. Installed this way onto water pipes, the pressure source (water) must be kept separate from the center fluid by an impermeable membrane.

The two most commonly used center fluids in differential manometers are water and mercury. Water is used as the fluid (colored for visibility) when the pressure differential is expected to be less than 1.25 psi. If the pressure difference were 1 psi, for instance, it would move the internal fluid in a water manometer 2.31 ft. Anything much larger than this would be unwieldy.

Mercury is most often used in manometers as the center fluid. It is 13.6 times heavier than water (S.G. = 13.6) and it takes that much more force to produce the same amount of movement as in a water manometer. Thus, with a significant pressure differential, the fluid in mercury manometers only moves a matter of inches. This is much more readable.

To calibrate a manometer, start with both ends connected to exactly the same pressure (or disconnect it completely) and let the center fluid ride to the middle of the tube, so that it rises to the same level on both sides. Mark that as the zero point on the calibrated scale. Now connect it up to the two pressure sources, and read the rise or drop in fluid level. If the center liquid is mercury, inches of movement on the scale will be read. Convert it to feet. This is now <u>feet of mercury</u>. Convert it to <u>feet of water</u> (multiply by 13.6, the Specific Gravity of mercury). To solve for flow through the unit (if this is a venturi manometer), insert it into the formula for flow through venturi meters. This differential pressure reading, in feet of water, <u>is</u> (Ph_1 - Ph_2).

If the center fluid is water, read the differential, in inches. Convert it to feet of water. The Specific Gravity of water is 1, so there is no need to adjust for this. If the center fluid is another substance, multiply by the Specific Gravity of that substance.

PRESSURE DIFFERENTIAL METERS - OPEN CHANNEL FLOW

Weirs

To measure flow in sewers, streams, drains - open channels - weirs have become most popular. They are merely flat or notched metal obstructions with sharp top edges, inserted across the line of flow. The water builds up behind the height of the weir, it flows over, and the depth of the overflow water is measured. The depth of water going over the weir is referred to as the "head" on the weir. Since it will be deeper at greater flows, calculations can equate it to flow, and shortened formulas for flow over weirs of specific shapes have been developed.

The two most commonly used weir types are:

Rectangular Weir - for large flows.

$$Q = 3.33 \ L \ h^{1.5}$$

h = head on weir; it is <u>measured from edge of weir in contact with the water, up to the water surface.</u>

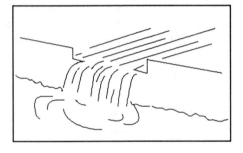

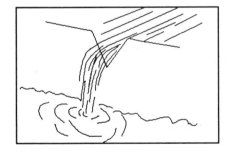

90 Degree V Notch Weir - for small flows

$$Q = 2.5 \ h^{2.5}$$

h = head on weir, and is <u>measured from bottom of notch to water surface.</u>

From a maintenance point of view, weirs may be thought of as the open channel equivalent of orifice meters; they have some of the same characteristics. Inexpensive to purchase or make of sheet metal with a wooden backup board for strength, they are easy to install and remove, are adaptable to a wide range of flows, and are very commonly used for temporary measurements of stormflows. They can be permanently installed, but solids carried by the water will catch on the lip or V notch, and decrease measurement accuracy. Frequent maintenance must be planned where weirs are in use.

A depth gage (calibrated stick) is usually used to take the head measurement. In permanent installations, electronic level sensors can be used. Water flowing over a weir contracts at the point of overflow, and the head measurement must be taken upstream of that contraction. Water flowing over a V notch weir must stay within the notch for accurate measurement. If it doesn't, a larger V notch, or a rectangular weir should be used. Water flowing over any weir must fall free of the weir plate (not dribble down the side) for good accuracy. If it doesn't fall free, a weir with a smaller notch should be used.

Weirs are often seen used in clarifiers, grit chambers, and other treatment plant units to reduce velocity or control depth in the unit. Saw toothed weirs, consisting of many small 90 degree V notches, are used as effluent structures in clarifiers. The flow leaving the clarifier is divided by the total number of notches on the weir to determine the discharge per notch. Head on the weir can be calculated from this. Installing a weir in a channel will effect deeper water (the height of the weir plus the head on the weir) immediately upstream of the weir.

When weirs are installed in sewers and manholes, this may cause backup problems, especially at times when stormwater enters the system. In addition, the extra depth created behind the weir produces a greater cross sectional area of flow; therefore, (Q = AV) velocity is decreased, in effect to zero, and solids will settle and collect here, throwing off measurement accuracy.

What happens downstream from the weir? Below is an exaggerated diagram of a sewer pipe with a weir installed. The slope of the channel is the same throughout.

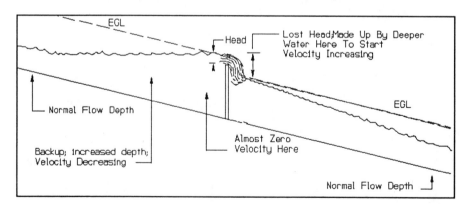

Parshall Flumes

The flume is similar to the venturi meter, except that it is meant to measure flow in open channels. The Parshall Flume is the most popularly used type, and is much used for the measurement of wastewater and irrigation water, due to its low permanent head loss, and self cleansing capacity. Its geometry is smooth flowing; solids will not get caught on it, and it operates well as both high and low flows.

The Parshall Flume has three sections:

Inlet Section - with rapidly converging walls and a level bottom. The decreasing diameter and lack of slope produces a deeper water here, and an optimal location for depth measurement for calculation.

Throat - the narrowest part, with a downward incline. Flumes are sized based on throat width (they range from 1 inch to 50 feet).

Outlet Section - diverging and sloping upward.

At low flows, water flows freely down the flume out. Formulas for calculation are based on this condition. At higher flows, a hydraulic jump forms at the outlet section, and as the flow increases, the jump will move upstream until it completely submerges the throat, restricting the flow.

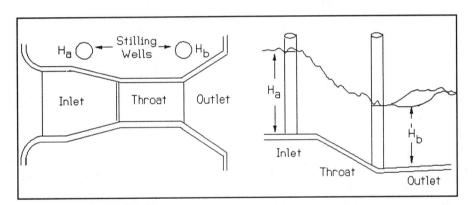

A depth measurement is taken at the first 1/3 of the inlet section with a calibrated stilling well (piezometer at the side of the flume), or one with a float gage, or by overhead ultrasonic unit. The entire depth of the water is measured. If flow is free and there is no submergence of the throat, depth at the well is measured and flow is estimated:

$$Q = 4 \ W \ H_a^{1.52} \ W^{.026}$$

W = width of throat

H_a = depth in stilling well upstream

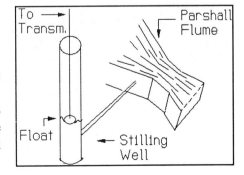

Formulas for flow through Parshall Flumes differ, dependent on throat width. The one above is used for widths of 1-8 feet, and applies to a medium range of flows.

If the downstream depth drowns out the jump and submerges the throat, the flow will be restricted, and the formula will not yield the true value. The correction depends on the percent submergence, and in flumes where this is a possibility, another stilling well is installed at the end of the throat section, so that this depth also can be measured. The percent submergence is obtained by dividing the downstream depth by the upstream depth, H_b / H_a , and the submergence graph below is referred to.

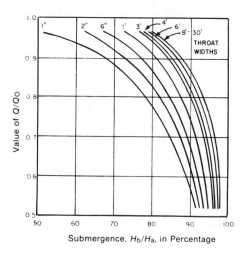

Submergence, H_b/H_a, in Percentage

Directions for Submergence Graph

Calculate flow from formula, using H_a for depth.
Find % submergence (H_b / H_a).
Enter graph at bottom; move up till proper throat width is reached, then left to Q/Q_o
    (a percent that the flow decreases by).
Multiply Q/Q_o by flow calculated from formula - yields corrected flow.

NOTE that only when the throat is more than 50-60% submerged will a correction be needed.

Other formulas for Parshall Flumes

Throat Width (W), in.	Flume Formula
1	$Q = 0.338 H^{1.55}$
2	$Q = 0.676 H^{1.55}$
3	$Q = 0.992 H^{1.547}$
6	$Q = 2.06 H^{1.58}$
9	$Q = 3.07 H^{1.53}$
12-96	$Q = 4 WH^{1.522 W^{8.874}}$

Maintenance on a Parshall Flume is minimal. However, it is important to keep the stilling wells free of debris.

The Parshall Flume is the most widely used type of flume for measuring flow, but it is not the only kind. The Palmer-Bowlus flume has had considerable application. Its advantages are that it is smaller in size, does not require a drop in the floor of the channel, and can be prefabricated from fiberglass, plastic, steel, etc. It is useful for temporary or permanent flow measurement at manholes.

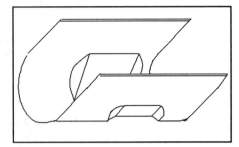

TRAJECTORY OF A JET

Flowmeters are not installed in every location where one might want to measure flow; fortunately, there are some simple methods of estimating quantity. For the locations where water freely exits the end of a pipe or channel, and drops to a surface below,

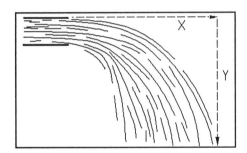

flow can be quickly computed based on measurement of the horizontal and vertical distance of travel after exit.

For these situations where we can measure the x and y axis of free fall we can easily calculate for flow:

$$Q = 4 \ A \ \frac{x}{\sqrt{y}}$$

Be aware that a common error is failing to note the area of flow. If the pipe is not full where the water exits, the Hydraulic Elements Curve will be needed to calculate area. If the channel is shaped other than round, a formula for the area of that shape will be needed.

These are specialized instances, but more common than one might think. Inplant, for recirculating flows, and for chemical feed, water often flows freely into a tank.

In the absence of flowmeters, discharge can also be estimated by filling or emptying a tank of known volume, and timing the process. For very small amounts, clock the amount of time it takes to fill a cylinder or bucket.

Derivation:

From basic physics, there is a well known formula which states that the distance of travel of a falling body, y, is equal to half its acceleration multiplied by the square of the time it falls:

$$Y = \frac{1}{2} \ a \ t^2$$

In addition, a body traveling horizontally, will travel a distance, x, which is equal to its speed times the time it travels:

$$x = \text{Velocity times time} \ - \text{or} - \ x/V = t$$

In our case of water exiting the end of a pipe, these two formulas can be combined and simplified in order to create one which includes both x and y, the measurable quantities:

$$Y = \frac{1}{2} \ a \ t^2$$

Acceleration = acceleration due to gravity:

$$Y = \frac{1}{2} \ g \ t^2$$

Substitute for t (we can't measure this):

$$Y = \frac{1}{2} \ g \ \frac{x^2}{V^2}$$

Equate to V:

$$V^2 = \frac{1}{2} \ g \ \frac{x^2}{y}$$

Use value of g and combine:

$$V^2 = 16 \ \frac{x^2}{y}$$

Eliminate exponents:

$$V = 4 \ \frac{x}{\sqrt{y}}$$

Equate to Q:

$$Q = 4 \ A \ \frac{x}{\sqrt{y}}$$

PITOT GAGE (Pitot Tube)

Invented in the 1700's, this is a flowmeter which relies on a direct measurement of
Velocity Head, based on differential gage
readings, as explained in Chapter 5. The section
facing upstream senses Pressure Head and
Velocity Head.   The downstream side only
measures pressure head. Subtracting for Velocity
Head, this can easily be converted to a flow rate.
The Pitot Tube is used to measure system and
fire flow, and can be attached to fire hydrants in
water distribution systems.

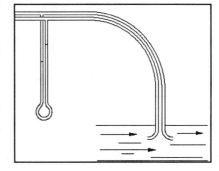

Fire Flow Testing

This is a practical application of pitot gage use: to determine fire flow or system head
conditions during fire flow, to predict pipe roughness and need for line cleaning, and
to assess risk of backflow conditions.

1. Connect pressure gage to a hydrant. This is the residual hydrant. All pressure
   readings are to be taken here, and it should be farther from the water source than
   the hydrant to be flowed, but on the same main.

2. One or more hydrants upstream are opened and flowed. Flow is measured using a
   pitot gage.

3. Measure pressure at residual hydrant before and during flowing. To get an accurate
   reading, make sure there is at least a 20 psi drop in pressure between static and flow
   conditions on the residual hydrant.

4. Solve for fire flow. This is one formula which can be used:

$$Q_f = Q_t \left(\frac{P_s - P_f}{P_s - P_t} \right)^{.54}$$

(use gpm & psi pressure)

Q_f = fire flow
Q_t = discharge during test (calculated from pitot)
P_s = residual pressure with no hydrants open
P_t = residual pressure during test
P_f = residual pressure during fire conditions (subtract minimum allowable pressure
      from P_s)

This variation works just as well:

$$\frac{Q_f \text{ (fire flow)}}{s^{.54} \text{ (under fire flow)}} = \frac{Q_t \text{ (test flow)}}{s^{.54} \text{ (under test conditions)}} \qquad \text{(use cfs \& ft. head)}$$

Cross multiply and solve for fire flow.

The formulas are used in the industry to obtain accurate fire flow capacity. A ratio is established between head losses at test conditions, and at fire flow conditions to obtain the maximum flow that can be achieved without dropping the pressure below minimum allowable level (usually 20 psi). It eliminates C factor and pipe diameter, the two values which are usually not exactly known.

MAGNETIC FLOW METER

In very common use in wastewater treatment plants, magmeters consist of a set of magnetic coils and an opposed pair of electrodes which are purchased as a flanged pipe unit and mounted in the process pipeline. The magnets are inserted onto crown and invert of pipe, and throw a magnetic field right through the pipe. The electrodes protrude through either side of pipe wall. It works on the principle that when an electrical conductor (the water) passes through an electromagnetic field, a voltage is induced at right

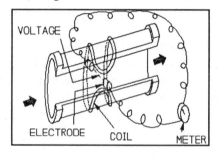

angles to the field, and this voltage is proportional to velocity. It is picked up by the electrodes, and then converted to a current signal, which can read out as a flow rate.

The magmeter offers no restriction to the flow, can be installed into short runs of pipe, is not affected by liquid properties. The electrodes must be kept free of solids.

ULTRA SONIC METERS

Recently developed, ultra sonic meters are being used to obtain flow readings in pipe systems, and in open channels.

For closed pipe flow, the transmissive type of meter sends sonic beams through the pipe from opposite transmitter/receivers mounted at a diagonal to the flowstream. The difference between upstream and downstream directed beams is measured, is directly proportional to velocity, and is calculated to flow. This meter is installed completely outside the pipe, and the pipe wall thickness must be exactly known. In use for clean water lines only, solids in the water will interfere with the flow reading. The reflective type of sonic meter is meant for wastewater use. A single sonic beam is sent into the water from a transducer mounted on the pipe; it bounces off solids in the water, and returns at a different frequency because of the water velocity. The magnitude of frequency change is measured, velocity is computed, and then flow.

For open channel flow - an overhead unit sends ultra sonic beams down to the water surface. This is often located at the head of a Parshall Flume in a wastewater treatment plant (where the stilling well would be). The beam bounces off solids in the water and returns to the sensor. Flow is determined based on the time it takes for the beam to return. It is actually measuring the depth of the water, and then calculating for flow. Restriction is that there must be solids in the water.

ROTAMETER

Rotameters are adaptable to small flows; they are accurate, but expensive, and are used

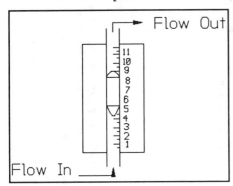

most often on chemical feed lines. This is a variable-area meter, whose operation is based on the continuity principle ($Q=AV$). The flow passes through a transparent, calibrated, diverging tube in which is suspended a float. The velocity of the water carries the float (heavier than water) up the tube until the area is wide enough to slow the velocity to the point where the float remains suspended and stationary. At that point one looks at the calibrations to read the correct flow.

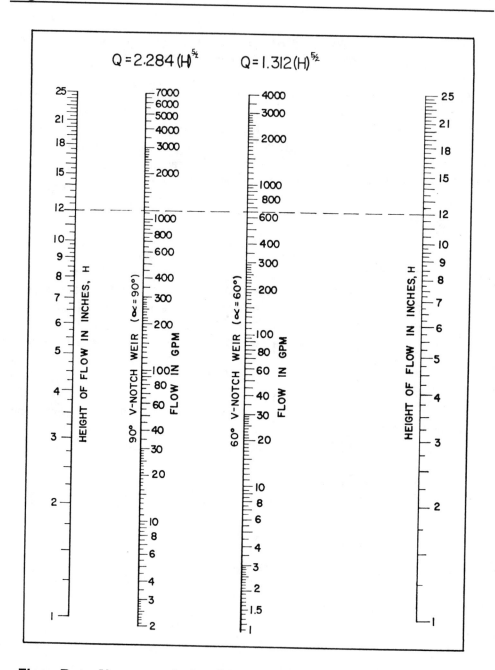

Flow Rate Nomograph for 60 and 90 Degree V-notch Weirs

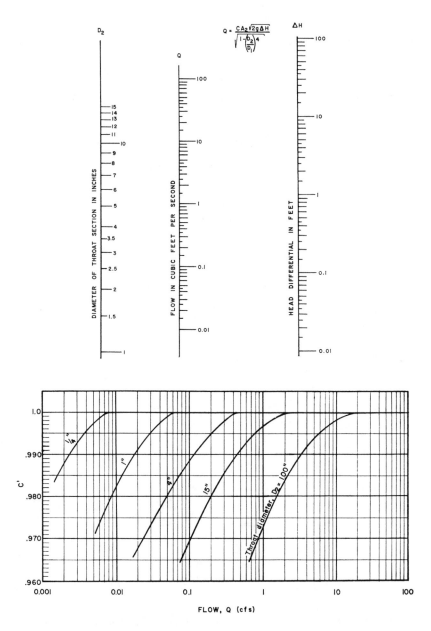

Nomograph and Corrections for Venturi Meter

Reprinted with permission of Public Works

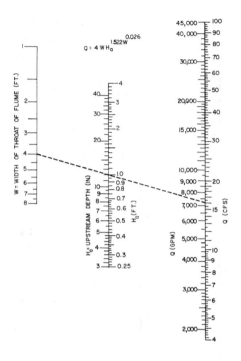

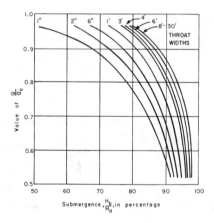

Nomograph and Corrections for Parshall Flume

Reprinted with permission of Public Works

PROBLEMS

1. Water passing through a 2 inch diameter new sharp edged orifice has a velocity of 22.5 ft./sec. and discharges to the atmosphere. What is the head causing flow?

2. Water exits a 4 inch diameter fire hose through a tubular nozzle with a single one half inch opening. If the pressure in the hose is 60 psi, what is the velocity of the water passing through this opening?

3. Corrosion creates a 1 inch diameter hole in the bottom of an elevated tank. The distance from ground to tank bottom is 60 feet. If C_d = .80, and 100,000 gpd is lost from this leak, what was the depth of the water in the tank when the leak started?

4. A 36 inch diameter water pipe with a pressure of 60 psi rubs against a rock and develops a 2 inch diameter hole in its side. How much water is lost in a day through this hole? (gpd)

5. A venturi meter reads 60 psi above the meter and 40 psi at the throat. Pipe diameter is 8 inches. Throat diameter is 2 inches. Calculate the flow (cfs)?

6. What should be the minimum weir height for measuring a flow of 1000 gpm with a 90 degree V notch weir, if the flow is now moving at 5 ft./sec. in a 2 ft. wide rectangular channel?

7. A weir 3 ft. high extends 20 ft. across a rectangular channel in which there is 100 cfs flowing. What is the water depth just upstream from the weir?

8. A 90 degree V notch weir is to be installed in a 24 inch diameter sewer to measure 500 gpm. What head should be expected?

9. A primary clarifier 30 ft. wide discharges .5 MGD over a weir the width of the tank. How high is the head on the weir (inches)?

10. What size orifice is required to discharge .565 cfs under a head of 28.5 ft.?

11. What is the flow through a 20 ft. by 16 ft. venturi meter with a 2 inch reading on a water differential manometer?

12. Water will exit a 20 ft. deep storage tank through a square cut hole to a flow-through process tank 4 ft. below. If the desired flow through the process is 5 cfs, what should be the dimensions of the hole?

13. A mercury manometer shows a 3 inch pressure difference across a 1 inch diameter orifice plate installed in a 6 inch line. If C_d = .68, what is the flow through this orifice?

14. A differential mercury manometer is installed on a pipe at an orifice meter. The differential reading across the orifice is 4 inches. What is the pressure difference in feet of water?

15. What is the gage difference on a mercury manometer, in inches, for a 3 cfs discharge through a 10 inch by 4 inch venturi meter?

16. A 6 inch by 2 inch venturi meter produces a 3 inch pressure differential on a mercury manometer.
    A. What is the pressure difference in feet of mercury?
    B. What is the pressure difference in feet of water?
    C. What is the flow through the venturi?
    D. What is the velocity of water at the throat?
    E. What is the velocity of water in the pipe upstream from the meter?

17. At the head works of a wastewater treatment plant, a Parshall Flume with a throat width of 3 ft. has a throat submergence of 18 inches and an upstream depth of 20 inches. What is the wastewater flow in cfs?

18. A 1 inch square hole develops in a blind flange. The pipe is 4 ft. off the ground. Pipe pressure is 80 psi. At what distance in front of the pipe will the ground get wet?

19. A 12 inch pipe flowing 1/3 full discharges from its end a stream of water which travels 2 ft. horizontally and 3 ft. vertically. How many gallons per day flow from the end of this pipe?

20. Water exiting the end of a hose at 1 cfs projects 3 ft. outward before hitting the ground 18 inches below. What is the diameter of the hose (inches)?

21. A tank 12 ft. long and 4 ft. wide contains 4 ft. of water. How many minutes will it take to lower the water to a 1 ft. depth if a 3 inch diameter plug (C_d = .6) is opened at the bottom of the tank?

22. A 48 inch diameter pipe with a flow of 2 cfs, a slope of .02 ft./ft., and a water depth of 6 inches, has a 90 degree V notch weir installed, whose height is 1.5 ft. to the bottom of the notch. For how many ft. will the water back up down the pipe?

23. An overflow masonry dam is to be constructed across a stream. Maximum flood flow is estimated to be 30,000 cfs. Six sluice gates 8 ft. high and 6 ft. wide (C_d = .85) are to be constructed in the dam (their sills at elevation 1122 ft.). The overflow weir will be 200 ft. long with a crest elevation of 1184. With all gates open under flood conditions, what will be the elevation of the water surface behind the dam?

24. A pitot tube in a pipe in which water is flowing is connected to a differential manometer containing water. If the difference in water levels in the manometer is 3.5 inches, what is the velocity of flow in the pipe (C_d = .9)?

25. A U tube which is open at both ends is partially filled with mercury. Water is poured into the left side of the U tube until the volume of water flows over the top. It is 2 ft. deep in the tube. Calculate the distance from the top of the U tube to the mercury in the right side of the tube.

26. Water flow for firefighting is being tested at a hydrant in a residential area. If during the test, pressure at the residual hydrant 300 ft. downstream is 60 psi static, and 43 psi under test conditions, what flow can be delivered for fire fighting at 20 psi if 1000 gpm was delivered during the test?

SOLUTIONS

1. **Answer**   21.6 ft. of water

   Sharp edged orifice C_d = .6; solve for flow; solve for head.

   $$Q = A \quad V$$

   $$Q = .785 \times .167^2 \times 22.5$$

   $$Q = .49 \text{ cfs}$$

   $$Q = C_d \quad A \quad \sqrt{(Ph_1 - Ph_2) \times 2g}$$

   $$.49 = .6 \times .785 \times .167^2 \quad \sqrt{(Ph_1 - Ph_2) \times 64.4}$$

   $$21.6 \text{ ft.} = Ph_1 - Ph_2$$

2. **Answer**   73.3 ft./sec.

   Short tube C_d = .8; change pressure to feet of water; solve for flow; then velocity.

   $$60 \text{ psi} \times 2.31 \text{ ft./psi} = 138.6 \text{ ft.}$$

   $$Q = C_d \quad A \quad \sqrt{(Ph_1 - Ph_2) \times 2g}$$

   $$Q = .8 \times .785 \times .0417^2 \quad \sqrt{(138.6 - 0) \times 64.4}$$

   $$Q = .1 \text{ cfs}$$

   $$Q = A \quad V$$

   $$.1 = .785 \times .0417^2 \quad V$$

   $$73.3 \text{ ft./sec.} = V$$

3. **Answer**   20 ft.

   Draw diagram. Hole in tank is an orifice; C_d = .8; solve for depth.

   $$.1 \text{ MGD} \times 1.55 \text{ cfs/MGD} = .155 \text{ cfs}$$

   $$Q = C_d \quad A \quad \sqrt{(Ph_1 - Ph_2) \times 2g}$$

   $$.155 = .8 \times .785 \times .083^2 \quad \sqrt{(Ph_1 - 0) \times 64.4}$$

   $$20 \text{ ft.} = Ph_1$$

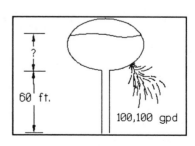

4. **Answer**   1,072,800 gpd

Change psi to feet of water; assume round edged orifice C_d = .8; solve for flow; change to gpd.

$$60 \text{ psi x } 2.31 \text{ ft./psi } = 138.6 \text{ ft.}$$

$$Q = C_d \quad A \quad \sqrt{(Ph_1 - Ph_2) \text{ x } 2g}$$

$$Q = .8 \quad .785 \text{ x } .167^2 \quad \sqrt{(138.6 - 0) \text{ x } 64.4}$$

$$Q = 1.66 \text{ cfs}$$

1.66 cfs x 60 sec./min. x 1440 min./day x 7.48 gal./cu.ft. = 1,072,800 gpd

---

5. **Answer**   1.17 cfs

Change pressure to feet of water; solve for flow.

$$60 \text{ psi x } 2.31 \text{ ft./psi } = 138.6 \text{ ft.}$$

$$40 \text{ psi x } 2.31 \text{ ft./psi } = 92.4 \text{ ft.}$$

$$Q = C_d \quad A \quad \sqrt{\frac{(Ph_1 - Ph_2) \text{ x } 2g}{1 - \left(\frac{d_2}{d_1}\right)^4}}$$

$$Q = .98 \text{ x } .785 \text{ x } .167^2 \sqrt{\frac{(138.6 - 92.4) \text{ x } 64.4}{1 - \left(\frac{.167}{.67}\right)^4}}$$

$$Q = 1.17 \text{ cfs}$$

---

6. **Answer**   1.2 ft.

Total weir height must be at least equal to depth of water plus head on weir. Change flow to cfs; solve for original depth; solve for head on weir; add.

$$\frac{1000 \text{ gpm}}{60 \text{ sec./min. x } 7.48 \text{ gal./cuft}} = 2.23 \text{ cfs}$$

$$Q = A \quad V$$

$$2.23 = 2 \text{ x d x } 5$$

$$.22 \text{ ft. } = \text{ depth}$$

$$Q = 2.5 \ h^{2.5}$$

$$2.23 = 2.5 \ h^{2.5}$$

$$.96 \ ft. = h$$

.22 ft. original depth + .96 ft. head on weir = 1.2 ft. tall (weir must be)

---

7. **Answer**   4.3 ft.

Solve for head on weir; add to weir height.

$$Q = 3.33 \ L \ h^{1.5}$$

$$100 = 3.33 \times 20 \times h^{1.5}$$

$$1.3 \ ft. = h$$

3 ft. height of weir + 1.3 ft. head on weir = 4.3 ft. depth.

---

8. **Answer**   .72 ft.

Change gpm to cfs; solve for head on weir.

$$\frac{500 \ gpm}{60 \ sec./min. \ \times \ 7.48 \ gal./cu.ft.} = 1.1 \ cfs$$

$$Q = 2.5 \ h^{2.5}$$

$$1.1 = 2.5 \ h^{2.5}$$

$$.72 \ ft. = h$$

---

9. **Answer**   .47 inches

Assume a flat weir; change MGD to cfs; solve for head on weir.

.5 MGD x 1.55 cfs/MGD = .775 cfs

$$Q = 3.33 \ L \ h^{1.5}$$

$$.775 = 3.33 \times 30 \times h^{1.5}$$

$$.039 \ ft. = h$$

.039 ft. x 12 in./ft. = .47 inches.

10. **Answer**   2 inches diameter

Assume sharp edged orifice $C_d = .65$; assume discharge to atmosphere; solve for diameter.

$$Q = C_d \quad A \quad \sqrt{(Ph_1 - Ph_2) \times 2g}$$

$$.565 = .65 \times .785 \ d^2 \ \sqrt{(28.5 - 0) \times 64.4}$$

$$.16 \ ft. = d$$

$$2 \ inches = d$$

---

11. **Answer**   840 cfs

Change inches water reading to ft. of water; solve for flow.

$$\frac{2 \ inches \ water \ (reading)}{12 \ in./ft.} = .167 \ ft. \ of \ water$$

$$Q = C_d \quad A \quad \sqrt{\frac{(Ph_1 - Ph_2) \times 2g}{1 - \left(\frac{d_2}{d_1}\right)^4}}$$

$$Q = .98 \times .785 \times 16^2 \ \sqrt{\frac{.167 \times 64.4}{1 - \left(\frac{.167}{.67}\right)^4}}$$

$$Q = 840 \ cfs$$

---

12. **Answer**   5.8 inches each side

Square orifice $C_d = .6$; solve for area of orifice, then side dimensions.

$$Q = C_d \quad A \quad \sqrt{(Ph_1 - Ph_2) \times 2g}$$

$$5 = .6 \quad A \quad \sqrt{(20 - 0) \times 64.4}$$

$$.23 \ sq.ft. = A$$

$$A = side^2$$

$$.23 = s^2$$

$$.482 \ ft. = s$$

$$5.8 \ in. = s$$

13. **Answer**   .054 cfs

Change orifice diameter to feet. Change inches of mercury to ft. of water. Solve for flow.

$$\frac{1 \text{ inch orifice}}{12 \text{ in./ft.}} = .083 \text{ ft.}$$

$$\frac{3 \text{ inches mercury}}{12 \text{ in./ft.}} = .25 \text{ feet of mercury}$$

.25 ft. mercury x 13.6 (SG mercury) = 3.4 ft. water

$$Q = C_d \quad A \quad \sqrt{(Ph_1 - Ph_2) \times 2g}$$

$$Q = .68 \times .785 \times .083^2 \quad \sqrt{3.4 \times 64.4}$$

$$Q = .054 \text{ cfs}$$

---

14. **Answer**   4.5 ft. of water

Change inches of mercury to feet of water.

$$\frac{4 \text{ inches mercury}}{12 \text{ in./ft.}} = .33 \text{ ft. mercury}$$

.33 ft. mercury x 13.6 (SG of mercury) = 4.5 feet of water

---

15. **Answer**   17.5 inches of mercury

Diameter change in this venturi is large, so formula for orifice meters can be used; Convert diameter to feet; solve for pressure differential; change to inches of mercury.

$$\frac{4 \text{ inches}}{12 \text{ in./ft.}} = .33 \text{ ft.}$$

$$Q = C_d \quad A \quad \sqrt{(Ph_1 - Ph_2) \times 2g}$$

$$3 = .98 \times .785 \times .33^2 \quad \sqrt{(Ph_1 - Ph_2) \times 64.4}$$

$$19.8 \text{ ft.} = Ph_1 - Ph_2$$

$$\frac{19.8 \text{ ft. of water}}{13.6 \text{ (SG mercury)}} = 1.46 \text{ ft. of mercury}$$

1.46 ft. mercury x 12 in./ft. = 17.5 inches of mercury

16. **Answer** A. .25 ft.
    B. 3.4 ft.
    C. .32 cfs
    D. 14.6 ft./sec.
    E. 1.6 ft./sec.

Diameter difference is large. Formula for orifice may be used.

A. Change inches of mercury to feet of mercury.

$$\frac{3 \text{ inches mercury}}{12 \text{ in./ft.}} = .25 \text{ ft. mercury}$$

B. Change feet of mercury to feet of water.

.25 ft. mercury x 13.6 (SG mercury) = 3.4 ft. of water

C. Change diameters to feet; solve for flow.

6 inches = .5 ft.

2 inches = .167 ft.

$$Q = C_d \; A \; \sqrt{(Ph_1 - Ph_2) \times 2g}$$

$$Q = .98 \times .785 \times .167^2 \; \sqrt{3.4 \times 64.4}$$

$$Q = .32 \text{ cfs}$$

D. Solve for velocity at throat.

$$Q = A \; V$$

$$.32 = .785 \times .167^2 \; V$$

$$14.6 \text{ ft./sec.} = V$$

E. Solve for velocity upstream.

$$Q = A \; V$$

$$.32 = .785 \times .5^2 \; V$$

$$1.63 \text{ ft./sec.} = V$$

17. **Answer**   21.6 cfs

Change upstream depth to feet; solve for flow; calculate % submergence; adjust flow.

$$\frac{20 \text{ inches}}{12 \text{ in./ft.}} = 1.67 \text{ ft.}$$

$$Q = 4 \ W \ H_a^{1.52} \ W^{.026}$$

$$Q = 4 \times 3 \times 1.67^{1.52} \times 3^{.026}$$

$$Q = 27 \text{ cfs}$$

$$\% \text{ submergence} = \frac{H_b}{H_a} = \frac{18}{20} \ .9 = 90\% \text{ submergence}$$

From % submergence graph, when throat is 90% submerged, flow is 80% of calculated flow.

27 cfs x .8 = 21.6 cfs

---

18. **Answer**   33 ft.

Assume $C_d = .6$; change inches to feet; change psi to feet; solve for flow; solve for horizontal distance of free fall.

1 inch = .083 ft.

80 psi x 2.31 ft./psi = 184.8 ft.

$$Q = C_d \ A \ \sqrt{(Ph_1 - Ph_2) \times 2g}$$

$$Q = .6 \times .785 \times .083^2 \ \sqrt{(184.8 - 0) \times 64.4}$$

$$Q = .36 \text{ cfs}$$

$$Q = \frac{4 \ A \ X}{\sqrt{y}}$$

$$.36 = \frac{4 \times .785 \times .083^2 \times X}{\sqrt{4}}$$

33 ft. = X

19. **Answer**   660,000 gpd

Pipe 1/3 full; solve for area (Hyd. Elements Curve); solve for flow; change to gpd.

From Hydraulic Elements Curve, pipe 1/3 full has an area which is 28% of the area of a full pipe.

$$A_{full} = .785 \; d^2$$

$$A_{full} = .785 \; x \; 1^2$$

$$A_{full} = .785 \; sq.ft.$$

$$.28 \; x \; .785 \; sq.ft. = .22 \; sq.ft. \; (area \; of \; our \; pipe)$$

$$Q = \frac{4 \; A \; X}{\sqrt{y}}$$

$$Q = \frac{4 \; x \; .22 \; x \; 2}{\sqrt{3}}$$

$$Q = 1.02 \; cfs$$

$$1.02 \; cfs \; x \; 60 \; sec./min. \; x \; 7.48 \; gal./cu.ft. = 660,000 \; gpd$$

---

20. **Answer**   4.3 inches

Assume hose is flowing full; change inches to feet; solve for diameter; change to inches.

$$18 \; inches = 1.5 \; ft.$$

$$Q = \frac{4 \; A \; X}{\sqrt{y}}$$

$$1 = \frac{4 \; x \; .785 \; d^2 \; x \; 3}{\sqrt{1.5}}$$

$$.36 \; ft. = d$$

$$4.3 \; inches = d$$

---

21. **Answer**   5.1 minutes

Draw diagram. Calculate volume of water to be removed. Solve for flow through the orifice; divide volume by flow to obtain time.

$$Vol. = l \; x \; w \; x \; h$$

$$Vol. = 12 \; x \; 4 \; x \; 3$$

Vol. = 144 cu.ft. This much water will be lost through the orifice.

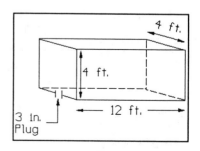

$$Q = C_d \quad A \quad \sqrt{(Ph_1 - Ph_2) \times 2g}$$

$$Q = .6 \times .785 \times .25^2 \quad \sqrt{(4 - 0) \times 64.4}$$

$$Q = .47 \text{ cfs}$$

$$\frac{144 \text{ cu.ft.}}{.47 \text{ cu.ft./sec.}} = 306 \text{ sec.} = 5.1 \text{ minutes}$$

---

22. **Answer**   95 ft.

Draw diagram. Solve for head on weir; add to weir height to notch for depth of water behind weir; subtract the depth of flow upstream of backup; use slope to calculate length of backup.

$$Q = 2.5 \quad h^{2.5}$$

$$2 = 2.5 \quad h^{2.5}$$

$$.9 \text{ ft.} = h$$

1.5 ft. weir ht. to notch + .9 ft. head = 2.4 ft. depth behind weir.

2.4 ft. depth behind weir - .5 ft. flow depth = 1.9 ft. backup depth

$$\frac{1.9 \text{ ft. backup depth}}{.02 \text{ ft./ft. slope}} = 95 \text{ ft. backup}$$

---

23. **Answer**   1191.6 ft.

Draw diagram. Solve for pressure on gates; solve for flow through gates; total flow for 6 gates; calculate amount going over the top; solve for depth (h on weir); add to original surface elevation; make adjustment.

1184 ft. (elev. of surface)
<u>1126 ft. (elev. of gate center)</u>
 58 ft. depth of water - pressure on gate

$$Q = C_d \quad A \quad \sqrt{(Ph_1 - Ph_2) \times 2g}$$

$$Q = .85 \times (6 \times 8) \quad \sqrt{58 \times 64.4}$$

$$Q = 2494 \text{ cfs . Flow through one gate}$$

2494 cfs x 6 gates = 14964 cfs through all gates

30,000 cfs (total flow) - 14964 cfs (flow through gates) = 15036 cfs over top

$Q = 3.33 \quad L \quad h^{1.5}$

$15036 = 3.33 \times 200 \times h^{1.5}$

8 ft. = h

1184 ft. (elev. of dam top) + 8 ft. head = 1192 ft. (elev. of water going over dam)

Stop! Once dam is in place and there is a head over it, the pressure on the gates is greater - and the flow through the gates is greater. Previous calculation is not accurate. Many adjustments would be needed for precision, but let's do it once.

1192 ft. (new top surface elevation)
<u>1126 ft. (center of gate)</u>
  66 ft. depth over gate

$Q = C_d \quad A \quad \sqrt{(Ph_1 - Ph_2) \times 2g}$

$Q = .85 \times 48 \quad \sqrt{66 \times 64.4}$

$Q = 2660$ cfs

2660 ft. x 6 gates = 15960 cfs through gates

30,000 - 15960 cfs = 14040 cfs over top

$Q = 3.33 \quad L \quad h^{1.5}$

$14040 = 3.33 \times 200 \times h^{1.5}$

7.6 ft. = h

1184 ft. (dam top) + 7.6 ft. head = 1191.6 ft. water surface over dam

---

24. **Answer**   3.96 ft./sec.

Solve for velocity; adjust for head loss.

Differential pressure = 3.5 inches = .3 ft. of water

Pitot tube reads Pressure Head at one gage and Pressure Head plus Velocity Head at the other. Differential pressure reading on manometer is equivalent to Velocity Head.

$$Vh = \frac{V^2}{2g}$$

$$.3 = \frac{V^2}{64.4}$$

$$4.4 \text{ ft./sec.} = V$$

Coefficient of discharge is a decimal number designating the % of loss of energy because of friction upon entering the unit. In the absence of a formula for pitot tubes, we can still use it by adjusting the velocity.

$$4.4 \text{ ft./sec.} \times .9 = 3.96 \text{ ft./sec.}$$

---

25. **Answer**   1.7 ft.

Draw diagram. Calculate movement of mercury; subtract from top.

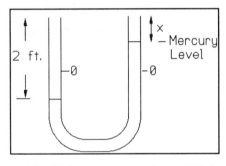

$$\frac{2 \text{ ft. of water pressure}}{13.6 \text{ (SG mercury)}} = .147 \text{ ft. (mercury moves this much up on right side)}$$

Before water was poured in, mercury rested at the zero point on both sides. On left side, water fills the tube from the top down to .147 ft. below the zero mark (because it moved the mercury out by this much). If the water extends to .147 ft. below the zero, and the mercury rose to .147 above the zero, then distance to the top of the tube on the right is:

$$2 - .147 - .147 = 1.7 \text{ ft. to top of tube}$$

---

26. **Answer**   2.28 MGD

Method #1

$$Q_f = Q_t \left(\frac{P_s - P_f}{P_s - P_t}\right)^{.54}$$

$$Q_f = 1000 \left(\frac{60 - 20}{60 - 43}\right)^{.54}$$

$$Q_f = 1587 \text{ gpm}$$

$$Q_f = 2.28 \text{ MGD}$$

Method #2

Head loss under fire conditions is 60 psi - 20 psi = 40 psi = 92.4 ft.
Slope for this head loss over 300 ft. is .308 ft./ft.

Head loss for test is 60 psi - 43 psi = 17 psi = 39.3 ft.
Slope for this head loss over 300 ft. of test space is .1309 ft./ft.

Test flow is 1000 gpm = 2.232 cfs

$$\frac{Q_f}{S_{fire}^{.54}} = \frac{Q_t}{S_{test}^{.54}}$$

$$\frac{Q_f}{.308^{.54}} = \frac{2.232}{.1309^{.54}}$$

$$Q_f = 3.5 \text{ cfs}$$

$$Q_f = 2.28 \text{ MGD}$$

CHAPTER 12

FLOW MEASUREMENT II TOTALIZER METERS

All the metering devices we have considered to this point have been flow rate meters, meant to provide a reading of cubic feet per second or gallons per minute, for utility process control use. This chapter will consider the totalizers; these meters count up the flow as it passes through the unit - for estimating water demand, and for billing.

POSITIVE DISPLACEMENT METERS

Referred to as Service Meters, positive displacement meters are suitable for residential customers who experience low flows and long periods when no water is used. The operating part is a measuring chamber which encloses a disc or piston. Each time a unit of water passes through the meter, the disc oscillates to pass the flow, and transmits the movement to a register, which records. The register unit is permanently sealed, and contains the register and a reducing gear train. A magnetic contact from the piston to the register unit drives it. Positive displacement meters are accurate at low flows, but have high head loss (approximately 5-10 psi), and are too expensive to use for mainline flow. They are normally used in 5/8 inch to 3 inch sizes, and normal flows should not be more than 1/3 of maximum meter capacity; operating them at excessive flows will result in wear and underregistration. Most residential services are metered with 5/8 inch meters that have 3/4 inch connections.

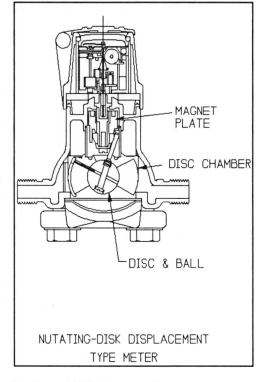

MAGNET PLATE

DISC CHAMBER

DISC & BALL

NUTATING-DISK DISPLACEMENT TYPE METER

TURBINE METERS

Turbine meters have a rotor in the measuring chamber which turns in response to the velocity of the water. There is either a direct linkage to the receiver, or the turbine blade may be fitted with a small magnet, and each revolution is picked up by a sensor on the outside of the meter. The movement is registered on a drive train which totalizes flow. Turbine meters are designed to handle large flows. The unit is inaccurate at low flows because small amounts of water slip past the rotor without moving it. Turbine meters generate very little head loss. The newer, more efficient models are called Turbo Meters, or Class I Turbines.

COMPOUND METERS

Also a customer meter, the Compound Meter is most useful where accuracy at both high and low flows is required. Within its body are a turbine meter on the main line, and a positive displacement meter on the bypass. At low flow, water passes through the positive displacement section, and the recording is of this section. The water continues on through the turbine section, but no recording takes place. When the pressure becomes great enough, the bypass shuts off and the turbine takes over, actuating its register. One of the disadvantages of this type of meter is the momentary loss of reading during the changeover from low to high flow. Designing the sections in series, as the one in this diagram, minimizes the loss, for the turbine is already in motion when flow increases enough for it to take over.

Compound Meters are the most expensive of the service meters, but if 10-30% of the flow is on the small side, the savings achieved in recording those low flows will probably be worth the cost. If not, a turbine should be installed.

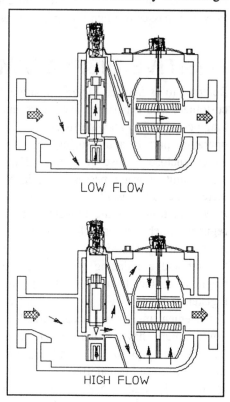

LOW FLOW

HIGH FLOW

PROPELLER METERS

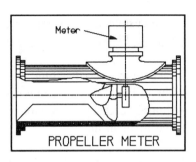

PROPELLER METER

Similar to the turbine meter, this unit is used in water mains where totalizing is required. Propeller meters are designed for constant high flow operation, and minimal pressure loss. The propeller is small, and mounted right in the main, facing upstream. Water velocity turns it and the register converts it to flow and totalizes.

PROPORTIONAL METERS

To measure fire flow water, a proportional meter is used. For testing of firefighting

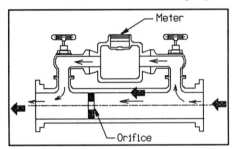

equipment, these are installed onto fire hoses. The metering apparatus is a small turbine or positive displacement unit installed into a loop which extends off the main line, and measures a small percentage of the water flowing through the line. A multiplying factor is built into the meter so that an accurate reading of the whole flow can be obtained. Fire line meters are designed this way because fire insurance companies require that no restrictions which could interfere with fire flow be installed onto the line (even if the meter were broken).

SIZING SERVICE LINES AND METERS

It is up to the discretion of the water utility to decide which and how many customers will be connected to the community system. Once connected, however, it is the responsibility of the utility to supply adequate pressure and flow to all customers, whether or not it is a metered system. There must be enough water to satisfy peak demand and fire flow, and pressures must be high enough throughout the system to minimize the possibility of backflow. The utility's responsibility extends up to the customer meter. It is not responsible for pressure losses within the customer's building; however, reasonable loss should be allowed for. When installing service to a new development or industry, the utility usually designs the service line so that there will remain in the line a pressure of at least 35 psi immediately downstream of the meter.

The type of demand must be considered. Industry usually requires a continuous flow. Irrigation water is the largest demand, but is of short duration. Domestic demand is the smallest, and use is intermittent.

FIRST: - Establish a peak flow demand (gpm needed) for the new establishment. Domestic water demand is due to the use of plumbing fixtures and washdown facilities by the customers. Assuming an average pressure of 35 psi available at the meter, the Plumbing Fixture Value Table lists various fixtures and a flow value for each.

--Take an average dwelling; refer to the table.
--Obtain Fixture Value (flow needed) for each fixture in the house.
--Multiply each by the number of fixtures of that type which will be in the house.
--Add them all up.

Now we have peak flow demand (gpm). This would be true if all residents of this dwelling use all these fixtures at once. Probability of this is slim. Therefore, make correction from Water-Flow Demand Per Fixture Value chart, and obtain a more realistic peak demand value for the dwelling. Multiply this value by the number of dwellings expected of this type.

--Do the same for other types of dwellings in the new area.
--Do it also for the industries.
--Add them all up. This is your peak demand. Use it as your flow.
--You may also want to adjust for projected future demand.

For an industry, often it is possible to start with the flow that this industry knows it will need, and eliminate this whole section for it.

NEXT: - Size the service line (lateral which runs from the main to the new establishment).

Information needed:

--Flow capability of main that serves the area.
--Static pressure in main during peak demand periods.
--Elevation difference between main and establishment.
--Length of service pipe from main to meter.
--Friction loss through meter and valves.
--Size of customer's piping and approximate head loss within building.

We have assumed that we need X amount of gpm and will want a 35 psi pressure.

--Measure elevation difference from main to establishment; convert to psi and add or subtract from main pressure.
--Draw picture of the new lateral with all fittings and valves. Measure length. Choose a diameter. If there is more than one pipe size involved, include it. Draw it out exactly the way it will be installed.
--Convert all fittings and valves to equivalent lengths of straight pipe. Add up. Add to pipe length.
--Using the peak demand flow just calculated, and the Hazen-Williams formula, obtain slope, then head loss. Convert to psi.

--Subtract the head loss from the static main pressure. Now we have the pressure at the end of the lateral, just upstream of the meter.

--If this is an industry, does it require an RP device? Subtract the psi lost from this.

Now - is there still at least 35 psi pressure left?

If not, choose a larger diameter pipe for the lateral, and try again.

There is one more thing to consider: At this flow and pipe diameter, calculate the velocity. It should be under 15 ft./sec. If is isn't, choose a larger lateral to slow down the water. The necessary flow will still be obtained, the pressure drop will be less, and hammer and corrosion will be minimized.

NEXT: - Choose a meter

Choose a meter of the type and size that is right for your flow and type of use. Meters should not be chosen by pipe size, or by pressure loss. Each meter type and size has been designed for a specific range of flows. AWWA has established standards for cold water meters based on flow capacities. The chart, Meters By Flow, lists various meters, their capacities and uses. The meter Meter Head Loss lists average pressure losses for different meters.

--A meter which is too large for the flow is a waste of money, and will let small flows slip through unrecorded.

--A meter which is too small for the flow which must pass through it will work too hard and wear out early.

--The meter chosen should have a maximum capacity of at least twice as much as the expected flow.

--Check the listed psi loss through the meter which is chosen. Subtract this from the static pressure in the lateral which has just been designed. If there is still 35 psi left, the expected flow will be delivered to the establishment.

Plumbing Fixture Value

Fixture Type	Fixture Value Based on 35 psi at Meter Outlet
Bathtub	8
Bedpan washer	10
Combination sink and tray	3
Dental unit	1
Dental lavatory	2
Drinking fountain (cooler)	1
Drinking fountain (public)	2
Kitchen sink: ½-in. connection	3
3/4-in. connection	7
Lavatory: 3/8-in. connection	2
½-in. connection	4
Laundry tray: ½-in. connection	3
3/4-in. connection	7
Shower head (shower only)	4
Service sink: ½-in. connection	3
3/4-in. connection	7
Urinal: Pedestal flush valve	35
Wall or stall	12
Trough (2-ft. unit)	2
Wash sink (each set of faucets)	4
Water closet: Flush valve	35
Tank type	3
Dishwasher: ½-in. connection	4
3/4-in. connection	10
Washing machine: ½-in. connection	5
3/4-in. connection	12
1-in. connection	25
Hose connections (wash down): ½-in.	6
3/4-in.	10
Hose (50-ft. length - wash down): ½-in.	6
5/8-in.	9
3/4-in.	12

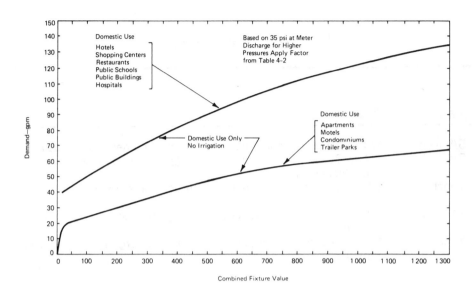

Water Flow Demand per Fixture Value - Low Range

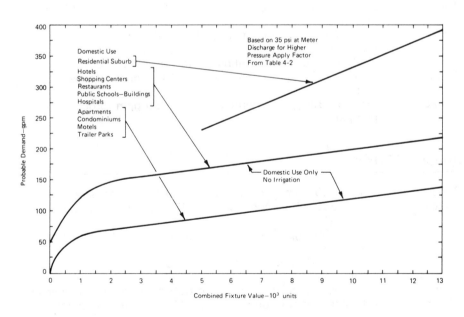

Water Flow Demand per Fixture Value - High Range

Meters By Flow

Displacement Meters

5/8 inch	1/8 to 20 gpm - continuous to 10 gpm (homes, small business)
3/4 inch	1/4 to 30 gpm - continuous to 15 gpm (homes, apartments, small business)
1 inch	3/8 to 50 gpm - continuous to 25 gpm (apartments, small business)
1 ½ inch	5/8 to 100 gpm - continuous to 50 gpm (motels, large apartments)
2 inch	1 ¼ to 160 gpm - continuous to 80 gpm (large hotels, housing complex)

Turbine Meters

2 inch	3 to 200 gpm - continuous to 160 gpm (hotels, housing complexes)
3 inch	4 to 450 gpm - continuous to 350 gpm (hotels, industry, irrigation)
4 inch	25 to 1000 gpm - continuous to 1000 gpm (large industry, irrigation)
6 inch	25 to 2500 gpm - continuous to 2000 gpm (processing, irrigation)
8 inch	140 to 1800 gpm - continuous to 900 gpm (processing, pump)
10 inch	225 to 2900 gpm - continuous to 1450 gpm (processing, pumps)
12 inch	400 to 4300 gpm - continuous to 2150 gpm (processing, pumps)

Compound Meters

2 inch	1/4 to 160 gpm - continuous to 160 gpm (motels, schools)
3 inch	1/2 to 350 gpm - continuous to 350 gpm (hotels, large hospitals)
4 inch	3/4 to 1000 gpm - continuous to 1000 gpm (apartments, dormitories)
6 inch	1 ½ to 1000 gpm - continuous to 500 gpm (apartment complexes)
8 inch	2 to 1600 gpm - continuous to 800 gpm (large condominiums)
10 inch	4 to 2300 gpm - continuous to 1150 gpm (large complexes)

Meter Head Loss

Displacement Meters

Size	Pressure Loss (psi at gpm)
5/8 inch	8.0 at 20
3/4 inch	10.0 at 30
1 inch	11.0 at 50
1½ inch	11.5 at 100
2 inch	12.0 at 160

Compound Meters

Size	Pressure Loss (psi at gpm)
2 inch	5.0 at 160
3 inch	4.0 at 320
4 inch	3.0 at 500
6 inch	14.0 at 1000
8 inch	14.5 at 1600

PROBLEMS

1. To prevent freezeup, a metered customer is allowed a let run at a rate of 2 gpm. If the customer is not to be billed for this water use, and the utility charges $.03/cubic foot registered on the meter for water, how much should be deducted from the customer's monthly bill while the let run continues?

2. Monthly water rates for customers of the city of Rockville are as follows:
   $1.00/1000 gallons for first 200,000 gallons used;
   $.80/1000 gallons for the next 500,000 gallons used;
   $.50/1000 gallons for everything in excess of this.
   If the meter totalizer at Acme Industries registers 500,000 cubic feet of water used this past month, what is Acme's water bill for this period?

3. A utility with 125 miles of water mains which produces an average flow of 12.12 MGD, has 8.71 MGD of metered customer sales, .37 MGD of metered use at public buildings, and 141 hours of firefighting/year at 2000 gpm. An average meter is underregistering by 2%.
   A. What is the percent lost water? (unaccounted for)
   B. Suppose unmetered public use is estimated at 1.6 MGD; recalculate lost water.

4. An unmetered water district has a nighttime flow of 1200 gpm and a daily flow of 2000 gpm. There is one industry within the district with a daily use of 800 gpm and a nighttime use of 600 gpm. There is a 40 ft. diameter water storage tank within the district in which the water level is rising at the rate of 2 ft. per hour at the time of the nightly flow. What is the nighttime domestic use?

5. A new trailer court is to be metered. The trailers will all be rental units, and there is maximum space for 130 units when the court is full. Expect 10 rentals the first year; one is in and occupied now. Expect many trailers to be occupied by senior citizens who will fly south for the winter. Fire protection must be provided. One meter installation will service the whole establishment. Nearest water main size is 12 inch diameter, with a service line extending to the court of 6 inch diameter. Required fire flow will be 1000 gpm. What meter installation will provide accurate flow readings for both high and low flow conditions for now, and for future years when the court is full?

6. An ice cube factory is moving into the city. It will need a 15 gpm flow for the ice cube machine, and 25 gpm for the water cooled compressors. There will also be a restroom and several hose connections. A 2 inch diameter service line services the building. A second ice cube machine may be installed in a few years. What meter should be supplied to meet present and future water needs?

7. The city golf club wishes to install an irrigation system, which will be metered. The service line size is 6 inch diameter, and it extends to a very small existing meter house. It is impossible to install a standard meter and backflow preventer and remain within the structure. The customer is very concerned about the pressure drops through the meter and RP device because sprinkler heads are at maximum spacing, requiring repumping to increase coverage area if pressures are reduced. The only length of straight pipe is a 9 ft. vertical pipe from the lower

level of the building. The lower level is not accessible. (Max. flow = 850 gpm; Min. flow = 100 gpm). What type of metering assembly should be installed to pick up the small flows (leaks), provide low head loss at high flows, and still fit in the confined space of the existing meter house?

8. An apartment building will be constructed (single story) to house 20 families, on level terrain, at a distance of 300 ft. from an 8 inch diameter supply main with a pressure of 80 psi. Size the service line to the apartment, and install the meter.

9. An electroplating outfit in Florida which required 200 gpm for the process is purchasing a building which is 200 ft. from the nearest main. The industry will require at least 50 ft. of 1.5 inch diameter piping inhouse for processing. It may expand in a year or so. It also has a bathroom. Main pressure is 65 psi. Size the service line to the building, and install the meter.

10. Read this meter.

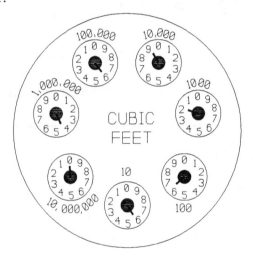

SOLUTIONS

1. **Answer** $346.52

Solve for gal./day used by let run; change to cu.ft.; solve for cost.

2 gal./min. x 1440 min./day x 30 days/mo. divided by 7.48 gal./cu.ft. = 11551 cu.ft./mo. water used

11551 cu.ft./mo. x $.03/cu.ft. = $346.52 deducted from bill.

2. **Answer** $2120.00

Change meter reading to gallons; total up cost of gallons used at each rate; add.

500,000 cu.ft. x 7.48 gal./cu.ft. = 3,740,000 gallons used

First 200,000 gallons:

@ $1.00/1000 gal. = $1.00 x 200 = $200.00

Next 500,000 gallons:

@ $.80/1000 gal. = .80 x 500 = $400.00

3,740,000 tot. gallons used
- 700,000 gallons already totalled
3,040,000 gallons left to be billed at $.50/1000

3,040,000 gal. (@$.50/1000 gal.) = $.50 x 3040 = $1520.00

Adding: $1,520.00 + $400.00 + $200.00 = $2,120.00 paid for water this month.

---

3. **Answer**   A. 26.2%
                 B. 13%

A. Add up accounted-for water; subtract from water made; obtain percentage.

fire water:

2000 gpm x 60 min.hr. x 141 hr./yr. = 16,920,000 gal./yr.

16,920,000 gal./yr. div. by 365 days/yr. = 46,356 gal./day = .05 MGD

underregistering meters:

8.71 MGD (customer meters) + .37 MGD (public meters) = 9.08 MGD

9.08 x .98 = 8.9 MGD actual metered water

8.90 MGD metered water
.05 MGD fire water
8.95 MGD accounted for

$$\frac{8.95 \text{ MGD accounted for}}{12.12 \text{ MGD total made}} = 73.8\% \text{ accounted for} \quad 26.2\% \text{ Lost Water}$$

B.   8.90 MGD metered water
                      1.60 MGD unmetered public
                       .05 MGD fire
                     10.55 MGD accounted for

$$\frac{10.55 \text{ MGD accounted for}}{12.12 \text{ MGD total made}} = 87\% \text{ accounted for} \qquad 13\% \text{ Lost Water}$$

Lost Water taken as water lost through leaks, not water lost for revenue.

---

4. **Answer**   286.8 gpm

Consider only nighttime use.

Water storage tank fills at rate of 2 ft./hr. Find gpm filling rate.

$$\text{Vol.} = .785 \times d^2 \times h$$

$$\text{Vol.} = .785 \times 40^2 \times 2$$

$$\text{Vol.} = 2512 \text{ cu.ft.}$$

$$\text{Vol.} = 18789.8 \text{ gal.}$$

Tank fills at a rate of 18789.8 gal./hr.

This is 313.2 gpm.

Water tank:

Rate pumped in - industry use - domestic use = Rate of fill

1200 gpm - 600 gpm - x = 313.2 gpm

x = 286.8 gpm Domestic Use

---

5. **Answer**   2 in. pos. displ. plus 6 in. turbine

Trailer Court

Limited use first few years plus senior vacancy in winter will mean low flows must be provided for. This is continual use, and revenue.

Fire flow needed at all times = 1000 gpm; only occasional use at the flow.

Domestic flow when full:

From Fixture Value Chart:
Bathtub	. . . . . 8 gpm
Shower	. . . . . 4 gpm
Water Closet . .	3 gpm
Dishwasher . . .	4 gpm
Washing Machine	5 gpm
Kitchen Sink . .	3 gpm

28 gpm x 130 trailers = 3640 gpm total

Adjusted value from Water-Flow demand chart = 3640 gpm, high range = 100 gpm

Full park domestic flow: 100 gpm

Present flow = 28 gpm -- adjusted 18 gpm daily use (less at night)

Fire flow = 1000 gpm

No one meter will service such high and low flows. A large meter would pass too much of the low flows. Install two meters - a 6 inch turbine on the 6 inch line, with a spring loaded check valve downstream which will not open until fire flows are encountered. On a bypass install a 2 inch positive displacement meter (records flows from 1.25 to 160 gpm) which will be in use most of the time.

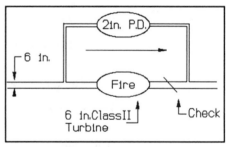

Note was made of winter vacancies. (cold weather). Install both meters in a pit to prevent freezing.

6. **Answer**   2 in. pos. displ. or 2 inch compound

Ice Cube Factory

Flow required:

Ice Cube Machines (2)	30 gpm
Compressors	25 gpm
Domestic:	
Sink	4 gpm
Water Closet	3 gpm
Hose Connections (4)	24 gpm
	31 gpm

Domestic adjusted value = 18 gpm

Total flow required with expansion:

30 gpm + 25 gpm + 18 gpm = 73 gpm

Ice cube machines will be on during the day only, as will domestic use, but compressors will be 24 hr./day use - lower flow at night.

Could use 2 inch positive displacement (records 1.25 to 160 gpm) or 2 inch compound meter (records .25 to 160 gpm).

Compound is more expensive, but catches all the low flows.

---

7. **Answer**   ?

Golf Club Irrigation System

RPPBP was needed because he already has booster pumping installed and intends to use when pressure is too low to get adequate water to far ends of course.

RP will drop pressure by 15-20 psi.

He wants a meter with low head loss. This knocks out the use of pos. displ., which would only be good for the low flows. Same with compound.-high head loss.

Turbine meter needs length of straight pipe upstream; this is not possible in this meter house.

This is an unusual situation. A right angle 6 inch propeller meter was installed (head loss = 1.5 ft.). Flow range 90-1200 gpm. This will not pick up the low flows; it will not register at all below 30 gpm.

The utility loses revenue.

What would you have installed?

---

8. **Answer**   2 inch diam. DIP service line - 2 inch pos. disp. meter

Apartment building - 20 units - obtain flow

From Fixture Value Chart:

Bathtub (20) @ 8 gpm . . . .   160 gpm
Water Closet (40) @ 3 gpm .   120 gpm
Shower (40) @ 4 gpm . . . .   160 gpm
Sink (40) @ 4 gpm . . . . . .   160 gpm
Dishwasher (20) @ 4 gpm . . .   80 gpm
Kitchen Sink (20) @ 3 gpm . .   60 gpm
Hose Connections (4) @ 10 gpm   40 gpm
Washing Machines (4) @ 12 gpm   48 gpm
Service Sinks (2) @ 7 gpm . . .   <u>14 gpm</u>
                                  842 total gpm

Adjust with Water-Flow Demand Chart (low range) -- 60 gpm practical flow needed

Draw diagram of system:

Will need <u>60 gpm (.133 cfs)</u> <u>at 35 psi.</u>

Elevation difference: Assume main 6 ft. below ground; allow 10 ft. different for elevation (4.3 psi).

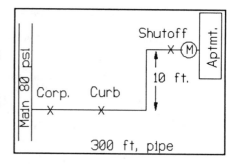

No RP device needed.

<u>Choose 2 inch diameter DIP (.167 ft. diam.) C=140</u>

Fittings:

3 open gate valves	4.5 ft. equiv. pipe
2 std. elbows	10 ft. equiv. pipe
straight pipe	<u>300 ft.</u>
	315 ft. total pipe length with fittings

Solve for HL_f:

$Q = .435 \quad C \quad d^{2/63} \quad s^{.54}$

$.133 = .435 \quad 140 \quad .167^{2.63} \quad s^{.54}$

$.0722 = s$

.0722 ft./ft. x 315 ft. pipe = 22.7 ft. HL_f = 9.9 psi

80 psi at main - 4.3 ft. (elev.) - 9.9 psi (HL_f) = 65.8 psi left at meter.

Check velocity at this flow:

$Q = A \quad V$

$.133 = .785 \times .167^2 \quad V$

6 ft./sec. = V

Velocity OK.

Continue. Choose a meter for the installation.

From Meters By Flow chart:

2 inch positive displacement will service up to 80 gpm max continuous demand.

This meter loses about 11 psi through the meter at 60 gpm.

Pressure 65.8 psi - 11 psi = 54.8 psi still left. Plenty of pressure, allows for some drop inside building.

---

9. **Answer**   6 inch service - 4 inch class II turbine

Required 200 gpm

Expansion 200 gpm - in Florida community it may easily double production.

Bathroom: (from fixture value table)
Sink . . . . 4 gpm
Water Closet . . . . 3 gpm

Also needed:
Safety Shower . . . . 4 gpm
Eyewash . . . . 2 gpm
Service Sink . . . . <u>7 gpm</u>
                                    20 total gpm

Adjust from Water-Flow Demand per Fixture Value Chart: Yields 15 gpm for this domestic use.

400 gpm industrial use + 15 gpm domestic use = 415 gpm Total flow needed (.927 cfs)

Industry intends to install 1½ inch interior piping (.125 ft.)

Try this side for service piping. check velocity.

$Q = A \quad V$

$.927 = .785 \times .125^2 \quad V$

$76 \text{ ft./sec.} = V$

No good - way too fast

NOTE: Interior piping - if flow is not split up immediately upon entrance, 1½ inch piping is too small inhouse. Lots of surge control and antisplash devices needed. Head loss through 50-80 ft. of this would be over 600 ft.

Service Pipe:

Try 3½ inch pipe (.292 ft.) Check velocity.

$$Q = A \quad V$$

$$.927 = .785 \times .292^2 \quad V$$

14 ft./sec. $= V$

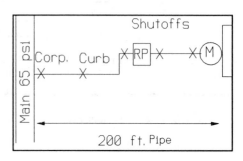

This will handle expansion flow and still have decent velocity. OK!

Draw diagram of system:

Fittings:
5 open gate valves . . . 10 ft. equiv. length
2 std. elbow . . . . 10 ft. equiv. length
Pipe length . . . . 200 ft.
                          220 ft. total length

65 psi at main

Elevation difference = 4 ft. (Fla. piping 1-2 ft. below ground) = 1.7 psi

RP device needed (electroplating) = 25 psi drop across it

Head Loss in service pipe: Choose new PVC: C=150

$$Q = .435 \quad C \quad d^{2.63} \quad s^{.54}$$

$$.927 = .435 \quad 150 \quad .292^{2.63} \quad s^{.54}$$

$$.153 = s$$

.153 ft./ft. x 220 ft. = 33.7 ft. $HL_f = 14.6$ psi

Subtract pressure losses:

65 psi - 1.7 psi (elev.) - 25 psi (RP device) - 14.6 psi (HL_f) = 23.7 psi at meter. <u>Not enough pressure!</u>

- This does not include any inhouse piping losses.
- It does not include the meter.
- Even at present flow of 215 gpm, the HL_f would be 10 ft., and pressure at the meter would drop to 34 psi - acceptable, but allows for no expansion.

A 6 inch diameter service pipe could be installed. HL_f would be:

$$Q = .435 \quad C \quad d^{2.63} \quad s^{.54}$$

$$.927 = .435 \quad 150 \quad .5^{2.63} \quad s^{.54}$$

$$.011 = s$$

.011 ft./ft. x 220 ft. = 2.4 ft. HL_f = 1 psi

Again: 65 psi - 1.7 psi - 25 psi - 1 psi = 37 psi at meter.

Choose a meter:

From Meters By Flow chart, a 4 inch Class II turbine would do. Max continuous use is 1000 gpm. Head loss is very low in turbine meters. Use would be during day, and at max. capacity all the time (no low flows).

This industry will probably still need booster pumping inhouse.

An alternate would be for them to have their own elevated water storage, a secondary supply, separated from community supply by air gap.

---

10. **Answer**   469,266 cubic feet

CHAPTER 13

CENTRIFUGAL PUMPS I

Man has always needed to move water from one place to another against the forces of nature. Gravity will move it downhill on a grade. If depth is built up behind, that pressure will move it further. But when pressure is dissipated by friction loss, and when water in the valley is needed on the mountain, the energy to move that water must be artificially created. We need a pump.

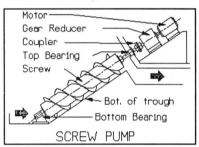

Archimedes invented the screw pump in 287 BC. It physically lifts volumes of liquid, but has space and height limitations.

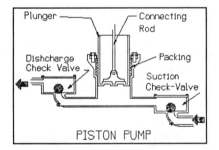

It is said that the Roman emperor, Nero, invented the piston pump at about 100 AD. Volume after volume of water is displaced with each stroke, but this is a high energy consumer, and size limits capacity.

Pumping technology was restricted to these, and variations, until the nineteenth century. The basic concept of the centrifugal pump existed during the 1600's, but development was slow because the high speed drives needed were not available for another 200 years. It wasn't until the 1800's that the first fully functional centrifugal pumps were developed. Since then, we have been able to move great volumes of water with much smaller units than with the pumps previously in use. Centrifugal pumps operate with high efficiency, few parts are required, initial cost is low, and maintenance is relatively easy.

Operation of the centrifugal pump is based on the principle that a high velocity is imparted to the water as it enters the pump, and that velocity is converted to pressure as the water exits the pump. From a mechanical standpoint, water enters the pump through the center, or "eye", of a rapidly rotating impeller. It is moved through the impeller vanes and thrown off their tips with great velocity. Still contained by the pump casing, the water cannot escape the unit, and collects at the "volute", or exit, slowing down as it leaves the pump.

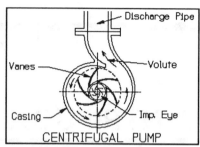

CENTRIFUGAL PUMP

From a hydraulic standpoint, note the energy changes that occur in the moving liquid. As water enters the pump from the suction piping, pressure is low. That's why the pump was installed. Passing from the suction piping through the impeller eye (smaller area), water velocity increases. This actually depresses the pressure even more, and acts to keep water coming in from behind, because of the relatively higher entrance pressure. As the water comes into contact with the rapidly spinning impeller vanes, velocity is increased

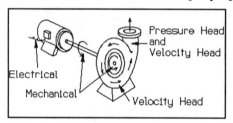

tremendously. This is the artificial energy which the mechanical action of the pump imparts to the water. It is Velocity Head energy which is increased at this point. The water is then rapidly thrown off the impeller tips against the pump casing, where it begins to slow down, collect, and then exit to the discharge piping through the only outlet, the progressively expanding volute. As Velocity Head decreases, Pressure Head increases; total energy must be maintained, and is merely converted from one form into another (Bernoulli's Theorem). The water now has enough new pressure to lift it the required distance.

The basic centrifugal pump working parts are impeller, casing, and the pump shaft, which connects it to the motor. There are no valves in the operating mechanism, and no close tolerances. Rotating parts do not touch stationary parts, and because of this, there is always some backward slippage of water between impeller and casing, which will vary depending on how hard the pump must work.

A CENTRIFUGAL PUMP DOES NOT ALWAYS PUMP THE SAME AMOUNT OF WATER.

This makes it very different from a piston pump. However, a smooth constant flow is produced, which does not pulsate, as in the piston pump.

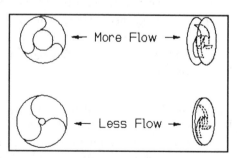

The capacity of a centrifugal pump is determined by the diameter of the impeller eye, and the width of the impeller vanes. The amount of pressure which can be created is controlled by the diameter of the impeller, and the rotating speed (velocity of the impeller rim). The faster it moves, the more water velocity is created, the more pressure will be available as it slows down at the exit.

Pumps with closed impellers are most efficient, and are used for clean water pumping. Pumps with semi-open impellers are used for solids handling water. Pumps with open impellers are sometimes used to pump sludges. Obviously the more open impellers have a better chance of passing solids without clogging.

PUMP TYPES - POSITION
Vertical Pump

Impeller shaft is in a vertical position. Well pumps are designed like this; impellers are at bottom of the well, and shaft extends upward to ground surface. Submersible pumps, with the motor close coupled to the pump, are vertical pumps used for underwater duty.

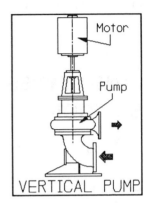

Horizontal Pump

Impeller shaft is in a horizontal position. The typical end suction pump is designed like this.

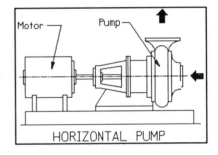

PUMP TYPES - EXIT FLOW DIRECTION
Radial Pump

Flow leaves the impeller at a 90 degree angle from the direction it enters the pump. Most centrifugal pumps are radial pumps. They are able to produce high pressures. The High Service Pump at a potable water treatment plant, which lifts water from the plant to elevated storage is usually a radial pump.

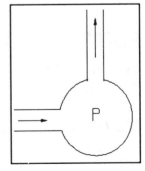

Axial Pump

Water enters and exits the pump on the same plane. Backwash pumps - and Low Service Pumps, which carry water from the source to the treatment plant, are usually axial pumps.

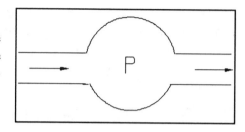

PUMP TYPES - SUCTION CONDITION

Single Suction

One suction line enters the pump casing.

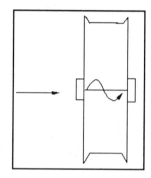

Double Suction

Suction line splits in two as it enters the pump and water approaches impeller from both sides. This type minimizes head losses inside the pump.

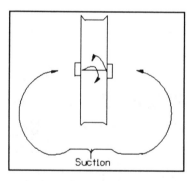

PUMP TYPES - IMPELLERS

Single Stage

One Impeller.

Multi-Stage

For uses where water must be delivered at high pressures (a high pressure pump). To feed boilers, membrane processes, and for deep wells. It is designed so that the discharge from one impeller is led through a passage in the casing to the suction of the next impeller.

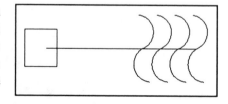

OTHER IMPORTANT PUMP COMPONENTS

Stuffing Box and Packing

Inspecting the diagram below, as the shaft extends outside the pump, water is prevented from escaping by the stuffing box, which is filled with rings of packing. The packing acts primarily as a seal, and also as a lubricant for the shaft. It is pressurized by an external source of water at a higher pressure than the water inside the pump. Often this is a small flow of the pump discharge water which is directed back to fill the packing. This water pressure is increased because it has already passed through the pump. Pump sealwater flow is small, but constant, and stuffing box is designed to allow a constant leakage.

Mechanical Seals

This is an alternate to sealwater and packing, and has two components:

Rotating Assembly - attached to the pump shaft.

Stationary Assembly - positioned permanently on the stuffing box.

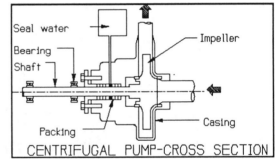

CENTRIFUGAL PUMP-CROSS SECTION

The face of the stationary assembly provides a running surface for the face of the rotating assembly. There is no water leakage from a mechanical seal.

Bearings

Bearings take up the stress of the rotating shaft. They hold it steady, and support the impeller. The rotating part of the bearing is locked to the shaft. The housing holds the stationary side of the bearing. There are two types, both of which are in most pumps. However, there is great variation in bearing position, type and number, depending on the pump.

Radial Bearing (line bearing) - this one is closest to the pump. It rides free in its own compartment and takes up and down stress.

Thrust Bearing - this one is farthest from the impeller, closest to the motor. It takes up the major thrust of the shaft, which is opposite from the discharge direction.

PUMPING CHARACTERISTICS:
THE PUMP PERFORMANCE CURVE

For a given centrifugal pump, the total pressure developed, the horsepower required to drive it, and the resulting efficiency, vary with the discharge. The hydraulic principles which we are familiar with in systems where the pressure source is an elevated tank of water, also apply to pumping. Pressure and flow vary with each other. A pump is installed because pressure is low or nonexistent. Its purpose is to add pressure to the system - to get the job done. This

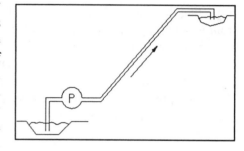

"head" which the pump must provide, or which it must "pump against", is dependent upon the requirements of the particular system. Viewing this typical diagram, the pump must lift water up the hill, and it must also overcome head loss in the piping while it is operating. <u>It must provide enough pressure to overcome lift and losses</u>.

<u>That is all the pressure the pump will provide</u>.

It will get the job done, and no more. It will meet the requirements of the system, to overcome the lift and losses, and get the water up the hill. By the time the water gets to the end of the pipe, pressure will be zero. Since every installation is different, yet not every pump is different, the flexibility is in the flow. If the lift up the hill is a great distance, or if piping head losses are great (or both), the pump will still do the work required; it will produce enough pressure, but will operate at a much smaller flow than if the lift and losses were smaller. The same is true when pumping a heavier liquid than water. If the Specific Gravity of the pumped liquid is greater than one, the pump will lift it the required distance, but the flow will be less than if it were pumping water. Horsepower requirements also vary with flow. Generally, if the flow is greater, the horsepower required to move the water will be greater.

How can a pump physically react with such flexibility? We should now recall the lack of close tolerances between the impeller and pump casing, which allows backwards slippage of water as the pump operates. This slippage is greater as the head (lift plus losses) increases, and therefore, flow decreases.

Pump performance is measured in gallons per minute delivered, height to which water is lifted, and efficiency. Each pump manufactured has a designated amount of head and flow at which it will operate most efficiently, and pump purchase should be based on how close to peak efficiency it will come in a particular system. This allows the pump to operate with the least strain, both mechanically and hydraulically. A pump which operates well off peak efficiency may do the job, but results will be excessive energy requirements and shortened pump life.

The interrelations of pump head, flow, efficiency and horsepower are known as the <u>Characteristics of the Pump</u>. These are important elements in pump performance, and they are diagrammed graphically on a Performance Curve. The curve puts these elements together to show a picture of what the pump can do. This is the Pump Characteristic Curve, and each centrifugal pump will be expected to perform as its curve describes. It will not stray from this.

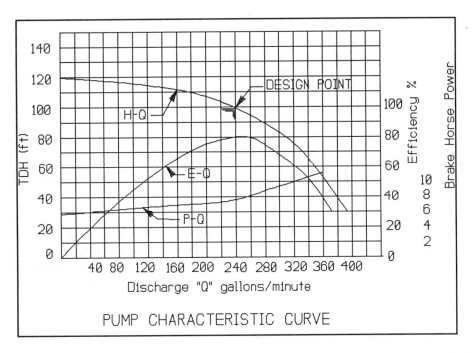

PUMP CHARACTERISTIC CURVE

If flow from a pump is measured, head (lift and losses) is calculated, and the characteristics do not match the curve, it is most likely that the losses have been estimated incorrectly.

Inspecting this typical Pump Characteristic Curve:

The horizontal scale on the bottom represents the flow in gallons per minute, which this pump will deliver. This is plotted against head, and creates the H-Q curve. This one is the base for most pump curves. The vertical scale represents head, in feet of water. Head is composed of lift plus losses which will be overcome by this pump. It is the total amount of useful energy that will be transferred from the impeller blades to the liquid. This is <u>Total Dynamic Head</u>, and it registers visibly as the difference between the suction and discharge gage pressures.

From the horsepower scale on the right is plotted power consumed, also against flow, and comprises the P-Q curve. This is <u>Brake Horsepower</u>, and refers us back to Chapter 6, where we first encountered power calculations:

$$\frac{gpm \times hd}{3960} = \frac{whp}{pump\ eff.} = bhp$$

The pump curve is actually derived from this formula. Try it. Pick any head on the curve. Read flow, and efficiency at that flow. Calculate brake horsepower with formula. Then look it up on the curve. It will match.

The Efficiency curve, E-Q, is also plotted against flow. Note that because of internal losses within the pump itself, no pump is 100% efficient. New pumps are usually about 85% efficient at best, and efficiency decreases with age.

Since all of the curves on the graph are plotted against flow, it is easiest when reading these curves, to determine flow first, then proceed up from the flow at the bottom to the curve with the element you desire.

The <u>Rated Head and Flow</u> are those at which maximum efficiency is reached. This is the <u>Design Point</u>, the point at which this pump was meant to operate. If we had purchased this pump for installation into our system, it would be because the head and flow conditions at this pump's design point matched the head and flow demands of our system.

Changing System Conditions

There is a maximum pressure that each pump can achieve. If the pressure needed is greater than that which the pump can deliver, the pump is said to be running at <u>Shutoff Head</u>. Shutoff Head is 120 ft., in the case of this pump.

For example, with a pump having this characteristic curve, installed into the system below, and assuming no friction losses, the pump would lift the water 120 ft. up the pipe. The pump would keep running, but the water level would remain there. The pressure of the discharge water in the pipe would be equivalent to the maximum pressure that this pump can achieve, and so there will be no flow. The water inside the pump will just circulate around and around, heating up, and will result in damage to the pump. This would be equivalent to operating the pump with the discharge valve closed. Horsepower requirements are lowest at this time; the water isn't going anywhere, just circulating around inside the pump casing.

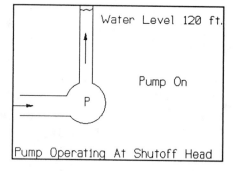

Pump Operating At Shutoff Head

It is, however, often desired to start up a centrifugal pump in this condition - with the discharge valve closed, when horsepower requirements are lowest, in order to start the flow moving slowly, then gradually opening the valve. This minimizes surge upon startup, and gives the water in the suction line a chance to overcome inertia and get moving into the pump.

Looking again at the characteristic curve, try extending the H-Q curve down farther than it is drawn, on the right, and you can imagine the pump running at high flow and negative head. This can happen upon starting a pump. A huge initial flow cannot overcome the inertia of the water in the suction pipe, and the water will split, or pull away from itself, creating a vacuum and drawing air into the pump. Centrifugal pumps have great difficulty expelling air, which may cause damage to the pump. It is better to start up with the discharge valve closed, or have the valve automatically operated so that it opens slowly as the pump starts. This phenomenon of high flow/negative pressure should not be new to us. The same thing occurs in a water line when flow suddenly increases, upon a main break, or a hydrant fully and suddenly opened. Large flow at once, vacuum created upstream, water separates. This invites a chance for backflow of cross connected contaminant into the system.

Let's take another example:

This pump is installed to discharge water horizontally through a pipe. There is no lift required, so the pumping head will be equivalent to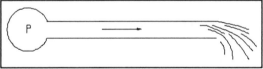
the head losses only. The flow delivered will be that which matches up on the characteristic curve with the system's head losses.

Now add a partially closed valve to the end of this pipe:

The flow delivered will be much less, because minor head loss created by the valve must be added to the friction head loss in the pipe. The
Total Dynamic Head to be achieved by the pump is greater; therefore flow will decrease. Refer back to Chapter 7 for Friction Losses in Fittings Chart. Note that head loss from a partially closed valve is significant, and in this particular system, this additional head loss is not very minor at all.

Now close that valve:

No water flows. The pump overheats. The same situation would exist if the valve were removed and the piping were extended to the point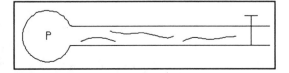

where the head losses in this horizontal line were greater than the pump could overcome at any flow. The pump would be operating at shutoff head, and the water would only go so far down the pipe, and then stop. For this situation we would need a pump with a higher shutoff head (a stronger pump), or we could install a booster pump midway down the pipe to raise the pressure again.

Another example:

If the head on the system should drop significantly for any reason (break in discharge line, perhaps), the pump would deliver a greater volume of liquid, and might easily reach the point of maximum horsepower requirement, and damage the motor.

If the needs of the system were expected to change in this manner, and a high flow were desired at intervals, from the same pump, then purchase of a larger pump would be in order, to fulfill this system requirement. To maintain best possible efficiency under all conditions, a pump with a flat H-Q curve would provide less variation in head over a greater capacity range.

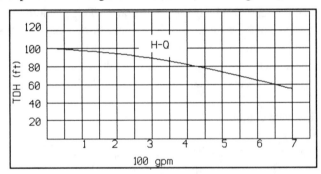

Sometimes changing to a wider impeller, or one with more vanes, will create the same effect.

If the condition of head-too-low, flow-too-high should exist at all times in a pumping installation, and be undesirable because this causes the pump to operate off peak efficiency, then running the pump with the discharge valve slightly closed will put more head loss into the system, and decrease the flow, bringing operation nearer top efficiency.

The System Curve

By now it should be apparent that in choosing a pump, we are trying to match the design point of a particular pump's curve with the conditions of our system, and if it does match, purchase that pump.

The system is the arrangement of pipe, fittings, and equipment through which the water will flow. The easiest way to match H-Q characteristics of the system with H-Q characteristics of various pump curves, is to develop a System Curve. The system head consists of lift plus losses. Lift is set, and can be measured, but head losses depend upon the amount of flow moving through the pipe. How can we determine the effect that the head loss of our system will have on a prospective pump's flow, when we can't calculate the head loss in the system until after we know the flow?

If we develop a system curve, with calculated heads based on a range of flows, then we will have the characteristics of our system, the actual conditions under which a pump will operate. Then we can pick the H-Q point at which we desire our pump to operate, and mark it. Now, inspecting several possible pump curves (representing several different pumps), we can superimpose each onto our system curve, and the one that matches our desired point, at a place on the pump H-Q curve which is at or near its design point - that is the pump which should be purchased.

To develop the system curve, choose a flow lower than that you wish to operate at, and calculate the head losses at that flow (Hazen-Williams); add this to the required lift, and mark it as "head" on the graph. Do this with several different increasing flows, and draw the graph from point 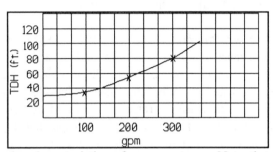 to point. This curve now represents the characteristics of your system. Note that system curves do not usually start at the bottom of the graph. The curve itself represents only the head losses. The level it starts at represents the lift.

If you wish to operate your system with more pressure in the lines than that which a pump will provide with the system as is, then theoretically add some more static head only to existing lift, and develop the graph with this included. Then after the pump is purchased and installed, a valve will have to be installed at the far end of the system to add this extra loss, and bring your flow into the range you had planned with the curve.

Don't forget that the system head includes both suction and discharge piping. Keep in mind that static head may vary, if dependent on tank drawdowns and filling. Both maximum and minimum static heads should be used to develop the system curve. The distance between them will be the actual operating range of the pump you choose. As short cut, head loss through pipe and fittings may be calculated for one flow, and losses for other flow rates determined from this relationship:

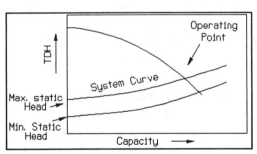

$$\left(\frac{Q1}{Q2} = \frac{H_1}{H_2}\right)^{.54}$$

Pumps in Series

When system heads are too great for one pump to overcome, pumps may be placed in series, one after another in the same pipeline. This puts more than one source of pressure into the system, and it <u>increases the head</u>. However, these pumps are pumping the same water through the same system. <u>Series pumping is not meant to increase the flow</u>.

We are familiar with:

Booster Pumping - the system has expanded; pipelines have been extended, or an apartment building has been added, and now more pressure is desired to lift the water that extra distance. A booster pump is installed at a low pressure area to raise the head to a safe and practical level for consumer use. When choosing a booster pump for the system, take the pump curves from original and booster. At the desired flow, add the heads of both to arrive at operating head with series arrangement. If the original

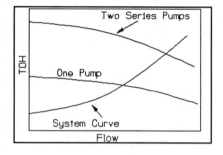

source of pressure is an elevated tower, this is still considered a series pump arrangement. Use the depth of water in the tower as the head for that unit. For a diagram of this effect, refer to the graph including the system curve transacting pump curve - with one pump installed, and then with two pumps installed.

Multi-Stage Pumps - a case of impellers in series in the same pump casing, hydraulically a series pumping setup. High pressure pumps, and deep well pumps must achieve great heads. The discharge of one impeller is directed into the eye of the next impeller. The flow is constant throughout, but at each stage the head increases. Each impeller is considered a separate pumping unit with a separate curve. Heads are additive, and total head is limited only by the efficiency of each. Identical impellers are used.

In most series pumping arrangements, identical or similar pumps are employed. To project horsepower use with pumps in series, use the total head to be achieved, and calculate. Efficiency can be calculated by multiplication of efficiencies of each.

Pumps in Parallel

When head is sufficient, but more flow is needed, pumps may be arranged in parallel. Wastewater lift station pumps, and treatment process pumps are placed in parallel in order to cope with the constantly fluctuating flow. We see parallel pumping also in industrial processing, where there exists specific needs for volumes of water for washing, rinsing, processing a product.

A diagrammatic sketch:

In this sketch the pumps are identical (though they need not be). From the standpoint

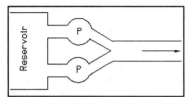

of each pump, the system requirements are the same, and must be met. Each pump achieves the same head for its flow, but since there are two of them, the system is getting twice as much water.

Pumps arranged in parallel increase the flow, but the head remains that of one pump working. When combining pump curves to determine characteristics, the combined discharge is the sum of the discharges of both pumps at that head. Once again, the curve with the system imposed should be referred to:

Parallel or series operations allow the operator to be flexible enough in pumping capacities and heads to meet requirements of system changes and extensions. Multiple pump operation also allows one pump to be taken out of operation for repair without shutting down the entire system.

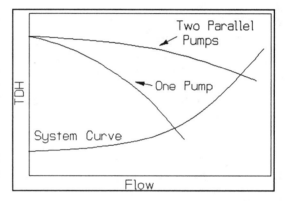

PROBLEMS

Many of the problems for this chapter refer to one of the five typical pump characteristic curves which are included following the list of problems.

1. Referring to pump curve #1:
    A. When the pump is operating under 90 ft. of head, what is the horsepower use?
    B. What is the head at which this pump will run most efficiently?
    C. What is the shutoff head for this pump?
    D. When this pump is operating at 30% efficiency, what head is it pumping against?

2. Referring to pump curve #2:
    A. At what two heads will this pump operate at 70% efficiency?
    B. This pump is designed to work at 130 ft. of head. Working at this design point, what flow is achieved?
    C. If the metered flow from this pump suddenly increases from 1200 gpm to 2200 gpm, what is the new head it is pumping against? What would you suspect caused this?

3. Given a pump with the characteristics of pump curve #1, what is the delivered flow under each of these conditions?
    A. Discharge lift 100 ft.; headloss in discharge piping 15 ft.; suction head 25 ft.; suction head losses 10 ft.
    B. Discharge lift 100 ft.; discharge piping head loss 35 ft.; suction lift 15 ft.; suction head loss 5 ft.
    C. Discharge lift 50 ft.; headloss in discharge piping 12 ft.; suction lift 20 ft.; suction head loss 13 ft.

4. Referring to pump curve #1:
    The pump has a static suction head of 15 ft., and is pumping 200 gpm from a tank. Total friction head loss in the system is 20 ft. What is the elevation difference between the pump and the end of the discharge line?

5. Referring to pump curve #1:
    A. What is the maximum percent efficiency this pump can achieve?
    B. What is the motor horsepower needed if the motor is 80% efficient and the pump is delivering 140 gpm.
    C. If the brake horsepower is 8 hp, what head is the pump achieving?

6. A pump delivers 100 gpm to an elevated storage tank whose bottom is 50 ft. above ground. Head losses in suction and discharge lines are 12 ft. total. Suction lift is 15 ft. The pump is 76% efficient and the motor efficiency is 85%. Power cost is $.15/Kw hr. What is the daily pumping cost?

7. A pump draws 60 Kw. Head loss is 9 ft. velocity is 5 ft./sec. Pipe is 3 inch diameter. If the wire to water efficiency is 60%, how high can the pump lift water in this system?

8. Below are the characteristics of a centrifugal pump:

Flow (gpm)	Total Head (ft.)	Efficiency (%)
200	124	30
400	123	54
600	119	67
800	114	76
1000	107	83
1200	98	84
1400	85	83
1600	73	76

Plot the curves showing the operating characteristics of this pump. What is the design point of this pump?

9. Construct a horsepower requirements curve for pump curve #4.

10. Referring to curve #5:
The electricity bill this past month was $5,000.00. The flowmeter on the pump records 800 gpm, and electrical cost is 10 cent/Kw hr. How efficient is the motor?

11. Referring to pump curve #5:
The pump lifts water 240 ft. and develops 27 ft. of friction head loss in the pipe. If the motor is 90% efficient, and electrical costs are $.08/Kw hr., find the cost to run this pump for a month under these conditions.

12. Referring to pump curve #1:
A. At what head will this pump draw minimal horsepower?
B. If the pump were operating with a suction lift of 12 ft., and discharging through 4000 ft. of horizontal line with a slope of .015, what would be the flow?

13. Water is pumped from a storage tank through a 4 inch diameter steel pipe. The velocity in the suction line is 6 ft./sec., and the distance between the pump and the storage tank is 20 ft. The pump discharges through a 3 inch diameter cast iron pipe 100 ft. long to an overhead tank. The water level in the overhead tank is 70 ft. above the level of the solution in the storage tank.
A. What is the pressure that the pump must develop? (psi)
B. If this pump is operating at 60% efficiency, what is the brake horsepower?

14. If a tank 150 ft. long, 10 ft. deep and 35 ft. wide is being dewatered with a pump rated at 100 gpm, and it takes exactly three days to empty the tank:
A. What is the actual pumping rate?
B. How efficient is the pump?

15. A wastewater pumping station has a static discharge head of 73 ft. The force main is 8 inch diameter DIP, 4600 ft. long. Minor station losses are estimated at 5 ft., and the flow rate is 600 gpm. What Total Dynamic Head are the pumps achieving? (C=120).

16. A pump delivers 1500 gpm through a 12 inch diameter pipe. How many 6 inch diameter pipes of the same length would be required to carry the same volume from the pump?

17. A low lift pump brings water from the lake to the treatment plant, which lies at the same elevation. System characteristics are:

    Suction Line:
    60 ft. of 4 inch diameter new DIP
    one open gate valve
    four 90 degree long sweep elbows
    reducer 4 inches to 2 inches

    Discharge Line:
    one open gate valve
    one open globe valve
    three 90 degree elbows - long sweep
    reducer 4 inches to 2 inches
    4 tees using through run
    1800 ft. of 4 inch diameter new DIP

    What is the water horsepower required to deliver 200 gpm to the treatment plant?

18. Water is pumped up 15 ft. from a tank through 30 ft. of 10 inch diameter suction piping with a gate valve and a standard elbow - to a pump which has the characteristics of curve #5, and then discharged through 500 ft. of the same diameter pipe past a check valve, a gate valve, two standard elbows into another tank 250 ft. above the pump. What is the flow from the pump?

19. Water flows through a main at a rate of 200 gpm. A pump with the characteristics of curve #1 is installed to take water from a tank and add to the flow in this main. The pump has a static suction lift of 10 ft., and a 4 inch diameter horizontal discharge line 200 ft. long. The main is 8 inches diameter and 3000 ft. long. What will be the flow from the pump?

20. A ductile iron pipeline 1000 ft. long will connect two reservoirs. One is 150 ft. higher than the other. A pump will be installed to pump water from the lower one up to the higher one at a rate of 500 gpm. If a pump with the characteristics of curve #2 is chosen, what should be the diameter of the pipe which connects the two reservoirs?

21. Given two identical pumps, each with characteristics shown in curve #2, what is the discharge for a head of 140 ft:
    A. If both pumps are operated in series
    B. If both pumps are operated in parallel

22. When a 3 stage well pump with characteristics as in curve #3 is in use - pumping from a depth of 1000 ft. through a 12 inch column - to elevated storage which is 80 ft. above ground - over an above ground discharge distance of 200 ft. What is the discharge of the pump?

23. The city is supplied with 1 MGD of potable water. Over the past few years, the city has increased population by 50%, and has expanded to the suburbs. Water storage was previously provided by a ground level tank; pressure from this tank is now inadequate for the outlying areas. An elevated tank 2 miles from the treatment plant will now be installed to replace it, and 80 psi will be available at the new tank.
    A. How high should the elevated tank be built?
    B. What horsepower high service pump will be needed?

24. An industry requires 400 gpm from a river 950 ft. away. The industry is 41 ft. higher in elevation than the river.
    A. Design the pumping system and size the pump to be purchased. Include fittings.
    B. If you had to choose from among the 5 pumps whose curves are given, which one would you choose?

25. A consulting engineering firm has been contracted to design a sewage lift station to move water into the treatment plant. The station must handle a 2 MGD flow. Pipe length is 700 ft. Vertical lift is 60 ft. Submersible pumps are desired. Must design for now and for the future. You are the consulting firm. Design it.

NO. 1

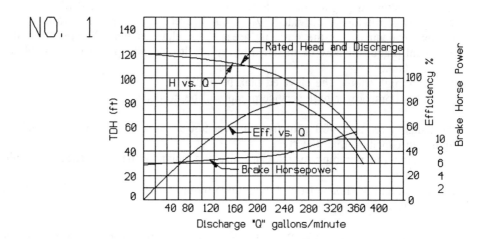

NO. 2

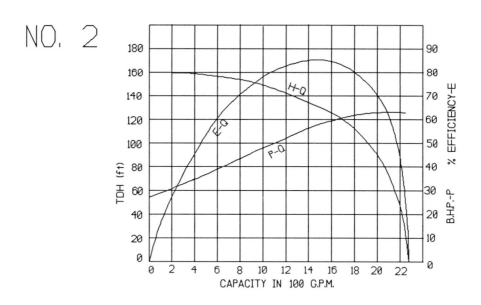

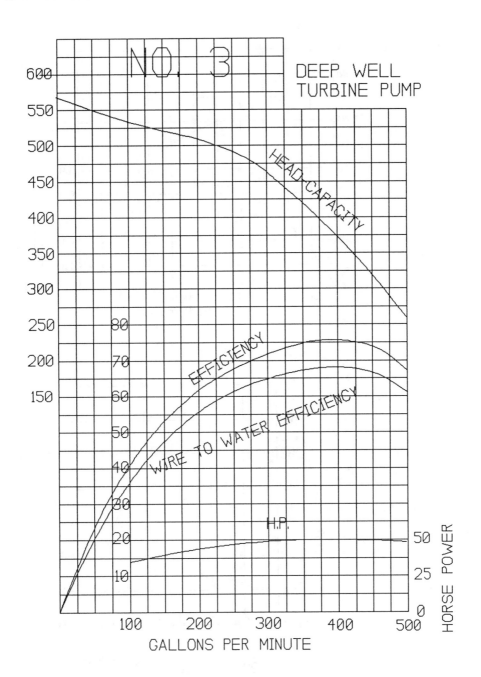

NO. 3 DEEP WELL TURBINE PUMP

GALLONS PER MINUTE

HORSE POWER

NO. 4

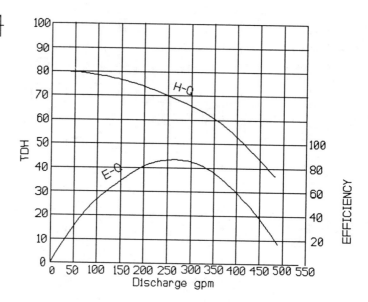

NO. 5

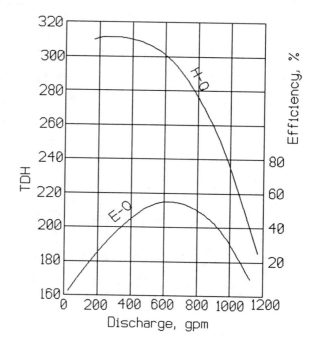

SOLUTIONS

1. **Answer**   A. 9 hp
                  B. 100 ft.
                  C. 120 ft.
                  D. 118 ft. and 46 ft.

To avoid mistakes, work from the flow.

A. On left, locate 90 ft. head; follow across to H-Q curve, then down to flow (276 gpm); now go back up to Hp curve, and across to the right to read hp = 9 hp.

B. Locate top of E-Q curve; slide down to the flow (240 gpm), then back up to the H-Q curve, then left to head = 100 ft.

C. Top of H-Q curve is shutoff head. Pump can lift the water only this high. Shutoff head = 100 ft.

D. This pump will be 30% efficient at two different heads.
    Locate 30% efficiency on left side of E-Q curve; slide down to flow (62 gpm); move straight up to H-Q curve, and read head on left = 118 ft. Now go to right side of E-Q curve (will have to extend the curves down a little to meet this point); slide down to flow (395 gpm), then back up to H-Q curve, and left to read head = 46 ft.

Note that if head is very high, it pumps little water and is very inefficient. Larger pump should have been purchased. If head is very low, a great deal of water is pumped, but also at low efficiency, for horsepower requirements are high.

---

2. **Answer**   A. 90 ft. & 155 ft.
                  B. 1500 gpm
                  C. 45 ft.

A. Locate E-Q curve; find point of 70% efficiency; slide down to flow, then up to H-Q curve; look left to TDH = 90 ft. Back to E-Q curve; find 70% efficiency on right side, slide down to flow, up to H-Q curve, over to head = 155 ft.

B. Locate head, slide down to flow = 1500 gpm

C. Locate new flow, up to H-Q curve, over to head = 45 ft. or less. Sudden great increase in flow and drop in discharge pressure indicates break in discharge line.

---

3. **Answer** A. 240 gpm
   B. 0
   C. 0

For each, draw diagram, solve for head; locate flow on curve.

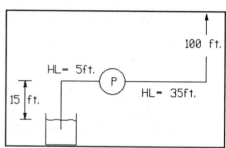

A. TDH = Lift + Losses

   TDH = (100 - 25) + (15 + 10)

   TDH = 100 ft.

   At this head the pump is operating at 240 gpm, the design point.

B. TDH = Lift + Losses

   TDH = (100 + 15) + 35 + 5)

   TDH = 155 ft.

   Head required is way over shutoff head. No flow delivered.

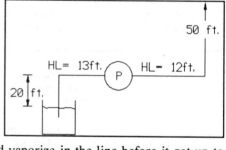

C. TDH = Lift + Losses

   TDH = (50 + 20) + (12 + 13)

   TDH = 95 ft.

   At 95 ft. of head pump delivers 260 gpm.

   Theoretically flow is 260 gpm. However, unless this is a specialized well pump, suction lift and losses are too great; water would vaporize in the line before it got up to the pump. There would be no flow.

4. **Answer** 103 ft.

Draw Diagram. Referring to curve, TDH at 200 gpm = 108 ft.; losses are 20 ft.

$$TDH = Lift + Losses$$

$$108 = x + 20$$

$$88 ft. = x$$

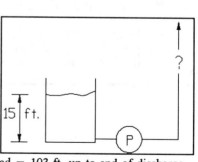

88 ft. lift + 15 ft. static suction head = 103 ft. up to end of discharge.

5. **Answer**   A. 80%
                  B. 7.5 mhp
                  C. 95 ft.

A. Refer to top of E-Q curve; look right for efficiency = 80%.

B. To find bhp, locate flow (140 gpm) on curve; slide up to P-Q curve; read bhp on right = 6 hp; divide by motor efficiency.

$$mph = \frac{bhp}{motor\ eff.} = \frac{6}{.8} = 7.5\ mhp$$

C. Find 8 hp on right; slide left to P-Q curve, then down to flow (260 gpm); At 260 gpm this pump uses 8 hp. Slide up from flow to H-Q curve; then left to head = 95 ft.

---

6. **Answer** $7.85

Draw Diagram. Solve for total head; then whp, then mhp, then cost.

TDH = Lift + Losses

TDH = (50 + 15) + 12

TDH = 77 ft.

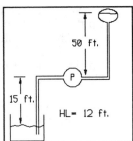

$$\frac{gpm \times hd}{3960} = whp$$

$$\frac{100 \times 77}{3960} = whp$$

1.9 hp = whp

wire to water eff. = .76 x .85 = .65

$$\frac{1.9\ whp}{.65} = 2.9\ mhp$$

2.9 mhp x .746 Kw/hp x 24 hr./day x $.15 Kw/hr. = $7.85/day

---

7. **Answer**   1719 ft.

Solve for whp; solve for flow; solve for head; solve for lift.

$$\frac{60\ Kw\ at\ motor}{.746\ Kw/Hp} = 80\ mhp$$

motor hp x wire to water eff. = whp

80 x .6 = 48 hp

$$Q = A \quad V$$

$$Q = .785 \times .25^2 \times 5$$

$$Q = .245 \text{ cfs}$$

$$Q = 110 \text{ gpm}$$

$$\frac{\text{gpm x hd}}{3960} = \text{whp}$$

$$\frac{110 \times \text{hd}}{3960} = 48$$

1728 ft. = hd. This is TDH (lift plus losses).

To obtain lift, subtract losses: 1728 ft. - 9 ft. = 1719 ft. lift.

---

8. **Answer**   98 ft. and 1200 gpm (84% efficiency)

Graph paper - along bottom set up range of flow (200-1600 gpm)
Up the side set up heads - up to 140 ft. at top, in 20 ft. increments.
For each flow stated, place dot at the corresponding head; draw H-Q curve.

Up right side set up efficiencies - up to 100%.
Plot each flow against efficiency the same way; draw E-Q curve.

On either side separately, set up horsepowers; calculate brake hp for each head, flow and efficiency. Plot each flow against its bhp on graph.

Choose the greatest efficiency this pump achieves; locate head and flow; this is the design point.

---

9. For each flow, head and efficiency, calculate bhp; plot this against flow on the graph; set up hp range on either side.

---

10. **Answer**   86%

Calculate mhp; refer to curve for head and calculate bhp; calculate motor eff.

mph x .746 Kw/hp x hr./day x days/mo. x rate = cost

mph x .746 x 24 x 30 x .10 = 5000

mhp = 93 hp

$$\frac{\text{gpm x hd}}{3960 \times \text{pump eff.}} = \text{bhp}$$

$$\frac{800 \times 276}{3960 \times .7} = \text{bhp}$$

80 hp = bhp

$$\frac{bhp}{mhp} = motor\ eff.$$

$$\frac{80}{93} = 86\% \ efficient$$

---

11. **Answer** $4,153.00

Add lift to Losses for TDH; refer to curve to calculate bhp; calculate cost.

240 ft. lift + 27 ft. HL = 267 ft. TDH

$$\frac{gpm\ x\ head}{3960\ x\ pump\ eff.} = bhp$$

$$\frac{860\ x\ 267}{3960\ x\ .6} = bhp$$

96.6 hp = bhp

96.6 hp x .746 Kw/hp x 24 hr./day x 30 days/mo. x $.08 = $4,153.00

---

12. **Answer**   A. 120 ft.
                  B. 325 gpm

A. At shutoff head it draws only 6 hp. It is also not pumping any water.

B. Solve for head; refer to curve for flow.

TDH = Lift + Losses

TDH = 12 + (.015 x 4000)

TDH = 72 ft.

At 72 ft. of head, the pump will deliver 325 gpm.

---

13. **Answer**   A. 41 psi
                  B. 9.1 hp

Draw diagram.

A. In order to solve for tot. hd.; solve for flow; solve for suction and discharge head losses; add lift + losses for TDH.

Q = A V

Q = .785 x .33² x 6

Q = .513 cfs

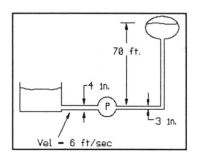

70 ft.

4 in.

P

3 in.

Vel = 6 ft/sec

Suction:

$Q = .435 \quad C \quad d^{2.63} \quad s^{.54}$

$.513 = .435 \quad 100 \quad .33^{2.63} \quad s^{.54}$

$.059 = s$

.059 ft./ft. x 20 ft. = 1.2 ft. head loss on suction side

Discharge:

$Q = .435 \quad C \quad d^{2.63} \quad s^{.54}$

$.513 = .435 \quad 100 \quad .25^{2.63} \quad s^{.54}$

$.229 = s$

.229 ft./ft. x 100 ft. = 23 ft. head loss on discharge side

TDH = Lift + Losses

TDH = 70 + (1.2 + 23)

TDH = 94.2 ft. head

$$\frac{94.2 \text{ ft.}}{2.31 \text{ ft./psi}} = 41 \text{ psi}$$

B. Convert flow to gpm; solve for hp.

.513 cfs x 60 x 7.48 = 230 gpm

$$\frac{\text{gpm x head}}{3960 \text{ x pump eff.}} = \text{bhp}$$

$$\frac{230 \times 94.2}{3960 \times .6} = 9.1 \text{ hp}$$

NOTE: Choice of C factor varies headloss.

14. **Answer**   A. 91 gpm
                  B. 91%

A. Obtain volume; solve for pumping rate.

Vol. = 1 x w x h x 7.48 gal./cu.ft.

Vol. = 150 x 10 x 35 x 7.48

Vol. = 392700 gallons

1440 min./day x 3 days = 4320 min. pumping time

$$\frac{392700 \text{ gallons}}{4320 \text{ min.}} = 91 \text{ gpm}$$

B. Divide by rated pumping rate.

$$\text{Eff} = \frac{91 \text{ gpm actual rate}}{100 \text{ gpm rated rate}} = .91 = 91\% \text{ efficient}$$

---

15. **Answer**   115 ft.

Solve for friction loss in force main; add to minor loss and static discharge head for TDH.

600 gpm x 1440 min./day = 864000 gpd = .864 MGD

.864 MGD x 1.55 cfs/MGD - 1.34 cfs

$Q = .435 \quad C \quad d^{2.63} \quad s^{.54}$

$1.34 = .435 \quad 120 \quad .67^{2.63} \quad s^{.54}$

$.008 = s$

.008 ft./ft. x 4600 ft. = 37 ft. friction loss

TDH = Lift + Losses

TDH = 73 + (37 + 5)

TDH = 115 ft.

---

16. **Answer**   6.4 pipes

Choose length and C factor; obtain head loss through 12 inch pipe at this flow; now obtain flow for a 6 inch pipe with the same characteristics; solve for number of pipes needed.

One foot of pipe chosen; C = 100

1500 gpm = 3.34 cfs

$Q = .435 \quad C \quad d^{2.63} \quad s^{.54}$

$3.34 = .435 \quad 100 \quad 1^{2.63} \quad s^{.54}$

$.008 = s$

$$Q = .435 \quad C \quad d^{2.63} \quad s^{.54}$$

$$Q = .435 \quad 100 \quad .5^{2.63} \quad .008^{.54}$$

$Q = .52$ cfs will pass through a 6 inch pipe

$$\frac{3.34 \text{ cfs}}{.52 \text{ cfs/pipe}} = 6.4 \text{ pipes}$$

---

17. **Answer**   2.6 hp

This pump is only required to provide for head losses. There is no lift. Convert fittings to length of straight pipe, add to pipe length; calculate slope, then head loss; calculate whp.

Suction line:

Length	60 ft.
1 gate open	2 ft.
4 90 degree ls els	28 ft.
d/D ½	4 ft.
	94 ft.

Discharge line:

Length	1800 ft.
1 gate open	2 ft.
1 globe open	100 ft.
3 90 degree ls els	21 ft.
d/D ½	6 ft.
4 tees	28 ft.
	1957 ft.

Total pipe length: 94 ft. + 1957 ft. = 2051 ft.

200 gpm = .45 cfs; new DIP C = 140

$$Q = .435 \quad C \quad d^{2.63} \quad s^{.54}$$

$$.45 = .435 \quad 140 \quad .33^{2.63} \quad s^{.54}$$

$$.025 = s$$

.025 ft./ft. x 2051 ft. = 51.3 ft. head loss in piping

Pumping head = 51.3 ft.

$$\frac{\text{gpm x hd}}{3960} = \text{whp}$$

$$\frac{200 \text{ x } 51.3}{3960} = 2.6 \text{ hp}$$

18. **Answer**   850 gpm

Draw Diagram. Must create system curve; obtain head at a range of flows; where the H-Q curve of #5 intersects, this is the operating point.

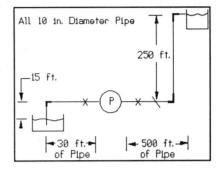

TDH = Lift + Losses

Lift = 15 ft. suction lift + 250 ft. discharge lift = 265 ft. lift.

Losses = HL for pipe length + fittings

Convert fittings to pipe length:

Suction side:

Straight pipe	30 ft.
Borda entrance	30 ft.
Std. elbow	25 ft.
1 gate open	6 ft.
	91 ft.

Discharge side:

Straight pipe	500 ft.
1 check	60 ft.
1 gate open	6 ft.
2 std. els	50 ft.
1 ord. exit	30 ft.
	646 ft.

91 ft. + 646 ft. = 737 ft. total length of pipe

Try 600 gpm (1.34 cfs)   New steel   C = 140

$$Q = .435 \quad C \quad d^{2.63} \quad s^{.54}$$

$$1.34 = .435 \quad 140 \quad .83^{2.63} \quad s^{.54}$$

$$.002 = s$$

.002 ft./ft. x 737 ft. pipe = 1.6 ft. HL

TDH = 265 ft. lift + 1.6 ft. loss = 266.6 ft. TDH

On curve #5, put an x at 600 gpm and 266 ft.

Try 800 gpm (1.78 cfs).

$$Q = .435 \quad C \quad d^{2.63} \quad s^{.54}$$

$$1.78 = .435 \quad 140 \quad .83^{2.63} \quad s^{.54}$$

$$.004 = s$$

.004 ft./ft. x 737 ft. pipe = 2.6 ft. HL

TDH = 265 ft. lift + 2.6 ft. loss = 267.6 ft. TDH

Put an x at 800 gpm and 268 ft.

Try 1000 gpm (2.23 cfs).

$$Q = .435 \quad C \quad d^{2.63} \quad s^{.54}$$

$$2.23 = .435 \quad 140 \quad .83^{2.63} \quad s^{.54}$$

$$.005 = s$$

.005 ft./ft. x 737 ft. pipe = 4 ft. HL

TDH = 265 ft. lift + 4 ft. loss = 269 ft. TDH

Put an x at 1000 gpm and 269 ft. head.

Draw system curve.

Pump curve crosses system curve at 850 gpm. This is the operating point of this pump.

---

19. **Answer**   350 gpm

Draw Diagram. Develop system curve for the combined pump system.

Flow will depend on lift and losses.

Flows are additive. TDH = Lift (10 ft.) + (loss in small line + loss in big pipe).

Obtain head for this pump at minimal flow; (get HL in small and large line; make sure to add flow together in large line). Try different larger flows till this system curve crosses the pump curve.

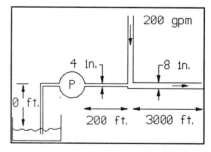

Try 50 gpm from pump through small line:

(.11 cfs) C = 120

$$Q = .435 \quad C \quad d^{2.63} \quad s^{.54}$$

$$.11 = .435 \quad 120 \quad .33^{2.63} \quad s^{.54}$$

$$.002 = s$$

.002 ft./ft. x 210 ft. pipe = .5 ft. HL in small line

Loss in large line:

$$Q = 200 \text{ gpm} + 50 \text{ gpm} = .56 \text{ cfs} \quad (C=100)$$

$$Q = .435 \quad C \quad d^{2.63} \quad s^{.54}$$

$$.56 = .435 \quad 100 \quad .67^{2.63} \quad s^{.54}$$

$$.002 = s$$

.002 ft./ft. x 3000 ft. = 6 ft. HL

Add head losses: 6 ft. + .5 ft. = 6.5 ft. Tot. HL if pump yields 50 gpm.

TDH = 10 ft. suction lift + 6.5 ft. HL = 16.5 ft. TDH

On curve #1, put an x at 50 gpm and 16.5 ft. head.

Try 200 gpm from pump through small line (.45 cfs).

$$Q = .435 \quad C \quad d^{2.63} \quad s^{.54}$$

$$.45 = .435 \quad 120 \quad .33^{2.63} \quad s^{.54}$$

$$.033 = s$$

.033 ft./ft. x 210 ft. pipe = 7 ft. HL in small line

Loss in large line: Q = 200 gpm + 200 gpm = 400 gpm (.89 cfs)

$$Q = .435 \quad C \quad d^{2.63} \quad s^{.54}$$

$$.89 = .435 \quad 100 \quad .67^{2.63} \quad s^{.54}$$

$$.005 = s$$

.005 ft./ft. x 3000 ft. = 15 ft. HL

Add head losses: 7 ft. + 15 ft. = 22 ft. Tot. HL if pump yields 200 gpm.

TDH = 10 ft. suction lift + 22 ft. HL = 32 ft. TDH

On the curve, put an x at 200 gpm and 32 ft. head.

Try 350 gpm from pump through small line (.78 cfs).

$Q = .435 \quad C \quad d^{2.63} \quad s^{.54}$

$.78 = .435 \quad 120 \quad .33^{2.63} \quad s^{.54}$

$.09 = s$

.09 ft. x 210 ft. = 19.3 ft. HL in small line

Loss in large line: Q = 350 gpm + 200 gpm = 550 gpm (1.23 cfs)

$Q = .435 \quad C \quad d^{2.63} \quad s^{.54}$

$1.23 = .435 \quad 100 \quad .67^{2.63} \quad s^{.54}$

$.0095 = s$

.0095 ft./ft. x 3000 ft. = 28.6 ft. HL

Add head losses: 19.3 ft. + 28.6 ft. = 48 ft. HL if pump yields 350 gpm.

TDH = 10 ft. suction lift = 48 ft. HL = 58 ft. TDH

On curve, put an x on 350 gpm and 58 ft. head.

System curve crosses pump curve here; no need to go any further. This is the operating point of the pump in this system.

NOTE: the lack of efficiency at this flow; this pump was a poor choice for this system.

20. **Answer**   8 inch diameter

Draw Diagram. Refer to curve for Tot. Head; subtract for head loss; solve for diameter.

From curve TDH = 158 ft.

150 ft. = Lift

8 ft. = Loss

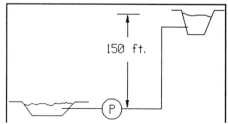

$$\frac{8 \text{ ft. head loss}}{1000 \text{ ft. pipe}} = .008 \text{ ft./ft. slope}$$

500 gpm = 1.1 cfs

DIP C = 130

$$Q = .435 \quad C \quad d^{2.63} \quad s^{.54}$$

$$1.1 = .435 \quad 140 \quad d^{2.63} \quad .008^{.54}$$

.585 ft. = d

7.1 in. = d

Choose 8 inch diameter pipe.

---

21. **Answer**   A. 1300 gpm
                  B. 2600 gpm

   A. When pumps are installed in series, the head increases; flow stays the same. If TDH = 140 ft., Q = 1300 gpm.

   B. Pumps installed in parallel draw from separate suction, and flows are addictive. Head is the TDH which corresponds to the flow - for each pump. At a head of 140 ft. the parallel system will produce a flow of 2600 gpm.

---

22. **Answer**   420 gpm

TDH = 1000 ft. to surface + 80 ft. above surface + Losses

Losses depend upon flow. Develop system curve.

Pump curve #3 represents one stage (1 impeller + bowl); this is a "pumps in series" problem (impellers in series). Heads are additive. Solve for head loss - at a range of flows. Where it crosses the pump curve, this is the operating point. Using three impellers will triple the head (at each flow, this pump will achieve three times the head of the same pump with one impeller). The three impellers for this system was necessary, for the lift alone requires it.

Solve for the head loss at each flow - it must be added to the lift to develop the system curve & TDH.

   Try 300 gpm: (.67 cfs)   Assume C = 100.

$$Q = .435 \quad C \quad d^{2.63} \quad s^{.54}$$

$$.67 = .435 \quad 100 \quad 1^{2.63} \quad s^{.54}$$

.0004 = d

.0004 ft./ft. x 1200 ft. pipe = .5 ft. HL

TDH = 1080 ft. + .5 ft. = 1080.5 ft. TDH

Try 500 gpm: (1.1 cfs)

$Q = .435 \quad C \quad d^{2.63} \quad s^{.54}$

$1.1 = .435 \quad 100 \quad 1^{2.63} \quad s^{.54}$

$.0011 = s$

.0011 ft./ft. x 1200 ft. = 1.3 ft. HL

TDH = 1080 + 1.3 = 1081.3 ft. TDH

To work this out accurately, a pump characteristic curve for the 3 stage pump is constructed. Take each flow (100,200,300,400,500 gpm), triple the head each achieves, and place an x at that point. Connect them to make the curve. This is the curve for the 3 stage pump.

Enter system curve on the graph. At no flow (static lift only) the TDH is 1080. Add head losses for each increasing amount of flow, and enter TDH's for these flows.

Head losses in this system are negligible. The system curve is very flat, and almost the total head needed is because of lift. We can divide the lift by three, neglect the head loss, and just extend the lift out to where it crosses curve #3 (H-Q). To adapt for use of this curve (impeller), obtain TDH (Total lift + Losses, then divide by 3. This yields the TDH that each impeller is achieving.)

---

23. **Answer**   A. 185 ft.
                  B. 80 hp

Assume 1.5 MGD will be an adequate supply for the expanded population. Assume city terrain is level.

A. 80 psi x 2.31 ft./psi = 185 ft. (Height of water surface in tank when full).

B.   **Pump:**

        1.5 MGD = 2.33 cfs = 1044 gpm

        TDH = Lift + Losses

        TDH = 185 + losses in main between treatment plant and tower.

        Find losses:

        2 miles = 10560 ft. of pipe

        Try 10 inch diameter main, new DIP (C = 140).

$$Q = .435 \quad C \quad d^{2.63} \quad s^{.54}$$

$$2.33 = .435 \quad 140 \quad .83^{2.63} \quad s^{.54}$$

$$.006 = s$$

.006 ft./ft. x 10560 ft. = 63.4 ft. HL (looks acceptable)

TDH = 185 ft. lift + 63.4 ft. loss = 248 ft. TDH

Assume high lift pump will be new and 85% efficient.

$$\frac{\text{gpm x head}}{3960 \text{ x pump eff.}} = \text{Bhp}$$

$$\frac{1044 \text{ x } 248}{3960 \text{ x } .85} = \text{Bhp}$$

$$77 \text{ hp} = \text{Bhp}$$

An 80 hp pump will be needed.

A smaller pump may be possible if a 12 inch diameter main is used. Cost of pump would have to be weighed against cost of pipe.

---

24. **Answer**   A. 7 hp
                   B. #4

Draw diagram.

A. Choose 6 inch diameter line; new steel pipe C = 140. Calculate equiv. pipe length for fittings; calculate slope, then HL; add to lift.

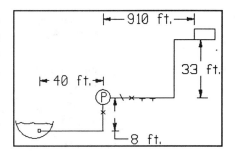

Suction Side:

Suction lift = 8 ft.

Pipe length	40 ft.
1 gate valve	3 ft.
1 foot valve	40 ft.
1 std. elbow	17 ft.
d/D ½	6 ft.
Borda entrance	18 ft.
	124 ft. Total pipe length
	Suction

Discharge Side:

Discharge lift = 33 ft.

Pipe length	910 ft.
d/D ½	10 ft.
2 std. elbows	34 ft.
1 check valve	30 ft.
1 gate valve	3 ft.
2 tees	20 ft.
	1007 ft. Total pipe length Discharge

124 ft. (suction length) + 1007 (discharge length) = 1131 ft. total length

Flow = 400 gpm = .89 cfs

$$Q = .435 \quad C \quad d^{2.63} \quad s^{.54}$$

$$.89 = .435 \quad 140 \quad .5^{2.63} \quad s^{.54}$$

$$.012 = s$$

.012 ft./ft. x 1131 ft. = 13.6 ft. HL

TDH = Lift + Losses

TDH = 41 + 13.6

TDH = 55 ft.

Calculate horsepower. Assume operation close to design point - new pump 85% efficient.

$$\frac{gpm \times head}{3960 \times pump\ eff.} = Bhp$$

$$\frac{400 \times 55}{3960 \times .85} = Bhp$$

6.5 = Bhp

Purchase a 7 hp pump whose design point is at this head and flow.

B. Inspect available curves.

#1 - would not provide enough flow at this head.

#2 - would have to have discharge valve closed almost to shutoff head in order to provide correct floor. Pump is too big. Power requirements too high.

#3 - This is a deep well pump and doesn't operate in this range at all.

#4 - This one has possibilities. At 53 ft. head it will deliver 400 gpm. However, efficiency is only about 60% at this point.

#5 - No good. Like #2, would have to be operating almost at shutoff head to provide this low a flow.

Will choose pump #4. Will have to keep head losses down in order to provide sufficient flow. 60% efficiency is poor, however. Would not purchase this pump as first choice.

---

25. **Answer**   Need 2 13 hp pumps which pump 1 MGD each
and 2 30 hp pumps which pump 2 MGD each

Draw Diagram.

2 MGD = 1389 gpm (3.1 cfs)

4 MGD = 2778 gpm (for future) (6.3 cfs)

Must be able to handle maximum flow with largest pump out of service.

Will need 2 small pumps, 1 MGD each - for low flow.

Will need 2 large pumps 2 MGD each - for high flow.

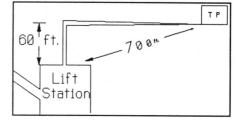

Bubbler for level control; signal relay for any 1, 2, or 3 pumps on.

TDH = lift (60 ft.) + losses (in 700 ft. of pipe)

Calculate head loss in 700 ft. line under worst condition (4 MGD flow - future)

Choose 12 inch diameter pipe; C = 130.

With 4 MGD flow:

$Q = .435 \quad C \quad d^{2.63} \quad s^{.54}$

$6.2 = .435 \quad 130 \quad 1^{2.63} \quad s^{.54}$

$.0166 = s$

.0166 ft./ft. x 700 ft. = 12 ft. HL in line at 4 MGD flow

With 2 MGD flow:

$Q = .435 \quad C \quad d^{2.63} \quad s^{.54}$

$3.1 = .435 \quad 130 \quad 1^{2.63} \quad s^{.54}$

$.005 = s$

.005 ft./ft. x 700 ft. = 3.2 ft. HL in line at 2 MGD flow.

Calculate pump size: Assume new pump 85% efficient.

Small Pump:
  (1 MGD); 694 gpm. TDH = 60 ft. lift + max 3 ft. loss = 63 ft.

$\dfrac{\text{gpm x head}}{3960 \times .85} = \text{Bhp}$

$\dfrac{694 \times 63}{3960 \times .85} = 12.9 \text{ hp}$

Large pump: (both together 4 MGD) 2778 gpm. TDH = 60 ft. lift + 12 ft. HL = 72 ft. Separately 2 MGD; 1389 gpm.

$\dfrac{\text{gpm x head}}{3960 \times .85} = \text{Bhp}$

$\dfrac{1389 \times 72}{3960 \times .85} = 30 \text{ hp}$

CHAPTER 14

CENTRIFUGAL PUMPS II

Once a pump has been installed, chances are that it will remain in operation whether or not performance is as expected. Pumps are sometimes purchased oversized, for instance, with expected duty to be heavier as system needs increase. A pump which is too powerful for the present system does not last longer. Pump curve will not match system curve and operation will be off maximum efficiency. On the other hand, a pump operating at design rating when first installed, will not be as efficient as system needs change with time.

CHANGING PUMP PERFORMANCE - PUMP AFFINITY LAWS

Up to this point we have considered only system changes and factors that will affect pump operation. However, the pump itself has some flexibility for change. Pump characteristics can often be adjusted to meet the system, and yet maintain peak efficiency. Each pump characteristic curve applies to that pump with the provided impeller, operating at the stated speed. Impeller size and pump speed are printed on the nameplate, but in many pumps they can be changed. Changing the rim speed of the impeller may be the most economical way to make adjustments. It will predictably affect head, flow and brake horsepower. This actually changes the rating of the pump and develops an entire new pump characteristic curve for it.

Pump speed may be changed in either of two ways:

Change pumping RPM - with a fixed gear drive - up or down, adjusting shaft speed. Extra space is required for this equipment.

Change impeller size - many pumps can be fitted with a number of different sized impellers. With a smaller impeller, for instance, the pump shaft is turning at the same rate, but the velocity at the rim of the impeller decreases. Not all pumps are originally purchased with the largest sized impeller installed. Many start with a smaller size, which better matches system characteristics at the time, and change to a larger impeller later, to adjust to the system.

The laws governing the effect of varying pump speed are known as <u>Pump Affinity Laws</u>, and are as follows:

When speed is changed, capacity varies directly as the speed, head varies directly as the square of the speed, and brake horsepower varies directly as the cube of the speed.

To simplify and explain:

$$\frac{New\ RPM}{Old\ RPM} \times Q = New\ Q$$

$$\left(\frac{New\ RPM}{Old\ RPM}\right)^2 \times Head = New\ Head$$

$$\left(\frac{New\ RPM}{Old\ RPM}\right)^3 \times bhp = New\ bhp$$

If impeller diameter is being changed, do the same thing (the effect is the same).

$$\frac{New\ Diameter}{Old\ Diameter} \times Q = New\ Q$$

$$\left(\frac{New\ Diameter}{Old\ Diameter}\right)^2 \times Head = New\ Head$$

$$\left(\frac{New\ Diameter}{Old\ Diameter}\right)^3 \times bhp = New\ bhp$$

Pump curves are available which demonstrate characteristics using different size impellers. They are called Oak Tree Curves (below) and are a way of demonstrating the performance range of the pump. The efficiencies appear on the curve like rings of an oak tree. The affinity laws show theoretical expected performance, but the curve shows exactly how the pump will operate.

Sometimes the width of impellers can be changed. A wider impeller throws a greater volume of water, and usually produces a flat Head-Capacity curve. Narrower impellers pump less and give a steeper curve.

Impellers with more vanes produce a flatter curve also. With less vanes a steeper curve is produced.

The size of the impeller eye can increase or decrease pump capacity.

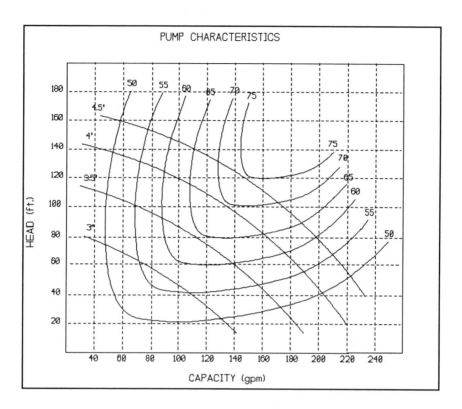

PUMP CHARACTERISTICS

VARIABLE SPEED PUMPS

For a system whose needs change with operation a variable speed pump may be the best choice. The speed of the pump will change to match the system needs. For instance, if the system must alternately deliver flow at different capacities, or if static lift is small, and head losses are great, and constantly changing, the pump head of a variable speed pump will vary with system head, and the pump chosen will operate efficiently at several different capacities.

Advantages of using variable speed pumps are in the energy savings obtained by operating close to peak efficiency regardless of changing system needs, and in the ability of the pump to discharge less water when demand is for less, with resultant smaller horsepower demand. These pumps are commonly used for wastewater pumping, especially at lift stations, where flows vary widely. They are good choices when the average quantity of liquid being pumped is less than two thirds of the maximum design flow. They are operated in response to information received from tank level controls; pump speed will be what is needed to keep the proper pressure differential.

Disadvantages of variable speed pumps are high cost, greater maintenance, and reduced overall efficiency.

PRIMING A PUMP

A major deficiency of centrifugal pumps is that the pump chamber must be filled completely with water upon startup in order for it to function correctly. If there exists a suction head (positive pressure on the suction side of the pump), the unit will always remain full, whether on or off, but with a suction lift, water tends to run back out of the pump and down the suction line when the pump stops. If the casing is filled with air or vapor, the impeller cannot create enough vacuum upon starting to draw water back into the unit, and the gas will just circulate around in the pump. The heat produced by the pump's mechanical action has no flowing water to dissipate it, and both pump and motor will overheat in a short time. It is best to always provide a positive suction head for a centrifugal pump, but if that is not possible, the pump must be separately primed (filled with water) each time it is started.

Rather than manually filling the pump with every startup, several different types of self-priming devices have been developed.

Foot Valve

A specialized check valve at the end of the suction line. It rides open when the pump is operating but shuts as the pump stops, trapping water in the unit and piping. Recorded efficiency of foot valves depends on installation; if the valve doesn't seat properly, water will leak out.

Vacuum Pumps and Ejectors

Separate units attached to the main pump casing, they create the vacuum needed to fill the pump before startup. These are commonly seen installed with low lift pumps at water treatment plants; disadvantage is extra capital cost and maintenance.

Priming Chamber

A permanently fitted reservoir of liquid above the pump. They are usually filled by a small recirculated flow of the pump discharge water, and empty their contents as the pump starts. Restriction is with size; this is a bulky unit, which must be able to completely fill the pump casing and associated piping.

Some self-priming pumps are built with a wide suction inlet which rides higher than the pump. Upon stopping, the discharge valve closes, and the water in this suction area slides back into the pump, rather than down the suction line. A vent exhausts any extra air.

CAVITATION

Water naturally contains a small gaseous phase, present as tiny bubbles or vapor molecules, which are in constant motion and always trying to escape the fluid. The percentage of this gaseous phase which remains in the water without evaporating, depends on the water's tendency to evaporate. At normal temperatures in an open vessel, atmospheric pressure upon the water minimizes this tendency; the vapor areas are fairly stable, and the water remains in the liquid phase. At elevated temperatures, or under conditions of lowered air pressure, the water's tendency toward vaporization increases. At higher elevations water boils at lower temperatures.

As water enters a pipe and pushes the air out, this evaporative tendency of the liquid, becomes, by comparison, more significant. If the water is under water pressure, instead of air pressure, the vaporizing tendency is minimal, but only because normal inpipe water pressures are usually a good deal more than atmospheric pressure (at 50 psi, water will not boil until a temperature of 300 degrees F is reached).

This tendency to vaporize, or, the pressure that the water exerts on its surroundings, is called <u>Vapor Pressure</u>. At 70 degrees F, water has a vapor pressure of .256 psia - very small. At 212 degrees F, water has a vapor pressure of 14.7 psia, the same as the atmospheric pressure upon it, and it boils.

When the pressure of the atmosphere falls below the vapor pressure of the water,the amount of vapor phase rapidly increases, bubbles are formed, and the water begins to flash into vapor. In a pipe, if the vapor cavity becomes so large as to fill the cross section of the pipe, a true column separation occurs. As we have seen, a water main break can cause this to happen. The increased flow creates vacuum, and some of the water flashes to vapor to fill the vacuum. If a valve is closed too rapidly, upstream of the valve pressure suddenly increases, causing hammer, but downstream of the valve, the Velocity Head keeps the water moving, vacuum forms in the space behind, and the water starts to vaporize in that area. Rapidly opening a valve can have the same effect on the upstream end. Power failure to a pump also will cause negative pressure waves to be set up which will travel downstream and produce vapor formation.

In the above mentioned cases, the pressure changes are momentary; as pressure equalizes in the system, and returns to the area, the low pressure condition is eliminated. The column rejoins, and the vapor bubbles collapse, or implode - quite forcibly - against the surface of the surrounding metal, chipping and disintegrating that metal, bit by bit. This phenomenon is referred to as <u>Cavitation</u>.

The effects of cavitation are most damaging in areas where it is occurring repeatedly. In pumping operations employing a suction lift, there is no positive depth of water to provide an initial pressure to the water entering the pump, and it is here that cavitation is most significant.

As a pump impeller rapidly increases water velocity, pushing fluid out of the pump, a tendency toward vacuum is created in the water coming in; in effect, a space is left between molecules of incoming water, and pressure is low (Velocity Head increases, Pressure Head decreases). If the pump is lifting water from below, it is only atmospheric pressure on the intake reservoir which pushes the water in to fill the suction pipe and take up these spaces. Since atmospheric pressure at sea level and standard temperature is equivalent to 34 ft. of water pressure, we could picture that a pump feeding a tank whose suction water surface elevation is at the same level as the pump centerline, really has a positive suction head of 34 ft. of water, as absolute pressure. Theoretically, we could lower the suction pipe to 34 ft. below the pump, and still get liquid into the pump. This amount of absolute pressure needed to get the water into the pump is called Net Positive Suction Head (NPSH). It is the total amount of energy, in Absolute feet of water, at the pump centerline.

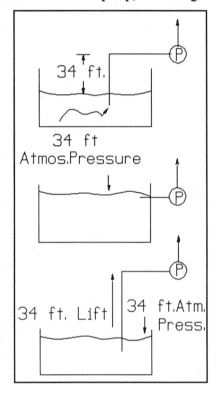

If this Net Positive Suction Head (theoretically 34 ft. or 14.7 psia) should drop below the vapor pressure of the water entering the pump, the external pressure will become less than the water's tendency to vaporize, and the water will start to flash into vapor at the pump entrance, eye, and impeller vanes. Bubbles will form, which will be forcibly collapsed as the water pressure suddenly increases near the pump exit. Cavitation will occur, with resulting chipping and destruction of internal pump parts. Operated this way for a length of time, the vibration caused will also damage bearings, shaft, and seals. The destructive effect is magnified further because of the fact that water molecules turning to vapor expand many times in the phase change. This enhances the repressurization and implosion.

But how does this theoretical 34 ft. of water pressure (14.7 psia) become less than the theoretical vapor pressure of water (.59 ft., .256 psia)? It does happen, because these are both only theoretical, and static quantities.

Consider first that there are friction losses within the mechanism of the pump itself. As the water enters the impeller eye, it makes a right angle turn. Then turbulence is increased by the vanes, adding to head loss. The eye, and the vanes, have a degree of roughness, etc. - all sources of head loss. Recognizing these internal pump losses, manufacturer's data will supply a specific amount of minimum positive absolute pressure which must be present in the water at pump entrance in order to overcome these head losses, and keep the pump running efficiently, and without cavitation. This is the Net Positive Suction Head Required (NPSHR). It is a characteristic of each pump, and is obtained by calculations and testing of the prototype pump of each model at a range of flows. It is determined by shape, size, and speed of pump; it increases with increased flow, and is often inserted as a NPSHR curve on the pump performance curve.

That is how much Absolute pressure is needed in order to use this pump. Now it must be determined how much pressure the system will provide. This amount is the Net Positive Suction Head Available (NPSHA), a characteristic of the system. It is defined as the energy actually available, in feet of Absolute Pressure, at the inlet of the pump. NPSHA is calculated for each installation with a simple formula:

NPSHA = (Atmospheric Pressure + Velocity Head) - (Lift + Losses + Vapor Pressure)

positive pressure forces   negative pressure forces

All of the elements of the formula are recorded in Absolute feet of water.

Atmospheric Pressure

All of the components of this formula can vary; atmospheric pressure may be 14.7 psia at sea level, but it is substantially lower in Denver, Colorado, a mile above sea level (see chart). Therefore, in a high altitude area, there is less positive pressure to start with.

WATER CHARACTERISTICS TABLE

Altitude Below or Above Sea Level	Atmospheric Pressure Pressure PSIA	Equivalent Head of 75° Water Feet (Abs)	Water Boiling Point °F
- 1000	15.2	25.2	231.8
- 500	15.0	34.7	212.9
0	14.7	34.0	212.9
+ 500	14.4	33.4	
1000	14.2	32.8	210.2
1500	13.9	32.2	209.3
2000	13.7	31.6	208.4
2500	13.4	31.0	
3000	13.2	30.5	
3500	12.9	29.9	
4000	12.7	29.4	204.7
5000	12.2	28.3	
6000	11.8	27.3	201.0
7000	11.3	26.2	
8000	10.9	25.2	197.4
9000	10.5	24.3	
10,000	10.1	23.4	193.7
15,000	8.3	19.2	184

Velocity Head

Velocity Head may be a small quantity, but when operating with a large suction lift, it may be significant enough to count. Use the water velocity at the impeller eye. This is the area of lowest pressure, and usually where the problem starts.

Suction Lift

Suction lift will vary with the installation, and also may vary with drawdown if water is taken from a tank. Use the lowest level.

Suction Losses

Suction losses will depend upon the flow, and will vary with it as the pump operates. Use the maximum flow the pump will be discharging at any time.

PROPERTIES OF WATER

Temperature °F	Absolute Vapor Pressure		Specific Weight lb./cu.ft.	Specific Gravity
	psia	ft. of water		
32	0.088	0.20	62.42	1.0016
40	0.122	0.28	62.43	1.0018
50	0.178	0.41	62.41	1.0015
60	0.256	0.59	62.37	1.0008
70	0.363	0.89	62.30	0.9998
80	0.507	1.2	62.22	0.9984
90	0.698	1.6	62.12	0.9968
100	0.949	2.2	62.00	0.9949
110	1.275	3.0	61.86	0.9927
120	1.693	3.9	61.71	0.9903
130	2.223	5.0	61.56	0.9878
140	2.889	6.8	61.38	0.9850
150	3.718	8.8	61.20	0.9821
160	4.741	11.2	61.01	0.9790
170	5.993	14.2	60.79	0.9755
180	7.511	17.8	60.57	0.9720
190	9.340	22.3	60.35	0.9684
200	11.526	27.6	60.13	0.9649
210	14.123	33.9	59.88	0.9609

NPSHA must be greater than NPSHR in order to prevent cavitation in the pump!

Vapor Pressure

Vapor pressure varies with water temperature. For cold water flow efficiently pumped, we can use .59 ft. as the vapor pressure. Note, however, that pumped water does absorb and dissipate heat energy created by the pump, and it does warm up. The higher the head, the lower the flow, the more the water heats up. Operating a pump with the discharge valve throttled almost to the point of shutoff head may heat up the unit enough to cause cavitation. Reducing flow by recirculating a percentage of discharge water back to the suction reservoir instead of throttling a valve may help, but often presents the same problem. The same water is being pumped over and over again, heating up more with each round trip. Hot climates, overheated pumphouses, direct sunlight, all contribute to the problem, increasing vapor pressure.

A cavitating pump usually makes noise like rolling marbles, gravel grinding, or balloons popping. When excessive, it can be very loud and rasping. Sometimes it can occur with no noise at all, and perhaps is then most serious, for it goes undetected.

In summary, we have discussed some of the factors which may drop NPSHA to below NPSHR, and start a pump cavitating:

> Excessive Lift
> Excessive Head Losses
> High NPSHR
> Low Atmospheric Pressure
> Hot Water

Other Causes of Cavitation

There are other factors which may contribute to cavitation within a pump:

Entrained Air

By this is meant outside air coming into the system on the suction side. This isn't as damaging as vapor caused cavitation, because there isn't the expansion of vapor within the pump, but its presence must be observed rather than calculated, and it is sometimes more difficult to discover. Air can enter the pump through pump packing, suction valve packing, and from flanged joints and shaft sleeves. The velocity of the water moving through the pump's system does create a slight eductor effect, and can easily suck air into the system through an opening.

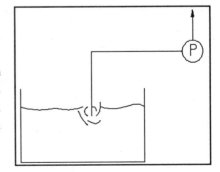

Insufficient Submersion

If the suction line is not submerged adequately in the suction tank, vortexing may be set up around the pipe, drawing air right in with the water.

Shallow Suction

A suction tank filling rapidly as the pump removes water, especially if it is waterfalling

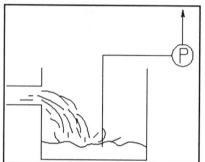

in, will trap excessive air, and may short circuit it right into the pump.

Slope of Associated Piping - if the discharge line is lower than the pump centerline, and the discharge valve is opened before the pump is started, vapor can escape backwards into the pump. The same can happen if the suction pipe is not designed with an inclined slope. Air pockets will form in the line, and as they build up, now and then flush into the pump. There should be 5-10 pipe diameters, or 3 impeller diameters from pump inlet port to nearest suction elbow. Increased turbulence at the elbow releases dissolved oxygen from the water, which may catch on the impeller. Highly aerated water will magnify this tendency.

At pump entrance the suction pipe decreases size through a cone shaped fitting called a reducer, and at discharge increases size again by the same method. Concentric reducers are only used on the discharge side. Eccentric reducers should always be used at suction entrances. When the supply line enters from below, an eccentric reducer with flat side up should be used at suction. It leaves no chance for air to collect in the pipe at pump entrance. When the supply line enters from above, (positive suction head), an eccentric reducer with curved side up should be used at suction. This gives any gas collected there a chance to slide back up the suction line away from the pump. Be aware that in this pumping arrangement (positive suction head), there is no risk of cavitation from low pressure conditions, but air pockets, no matter what the pressure condition, if becoming large enough inside a pump, can cause vapor lock, or air blockage. The air cannot be expelled and the pump's capacity will be greatly reduced (a pump

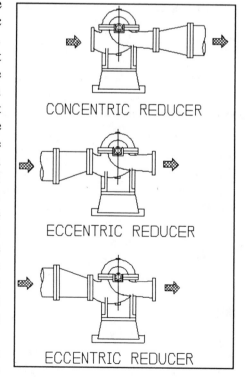

CONCENTRIC REDUCER

ECCENTRIC REDUCER

ECCENTRIC REDUCER

filled 5% with air will reduce capacity by 50%). If the air pocket is large enough, no water will be discharged at all. A pocket of air in a pump, a pipeline, a sand filter, or any other water-filled container, is usually not expelled by the moving water. Rather, the water routes around it, reducing the effective working volume of the unit.

In the above sections we have referred to air becoming entrained in the system from the outside, and to water vapor formations from within the liquid, but the same principles, and damage, can be applied to any gas which becomes trapped in pump or associated piping. In wastewater pumping it is likely that gases other than water vapor or air will be formed within the wastewater, and lead to cavitation. Wastewater pumps should always have a flooded suction, if possible.

CAVITATION IN PIPING

Cavitation can cause stress in piping when velocity is high and direction of flow is sharply changed. Water going around an elbow or tee very fast will pull to the outside

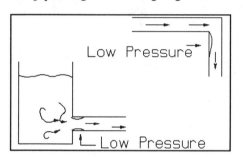

of the bend, creating thrust there, but on the inside of the bend, pressure may drop to a point corresponding to the vapor pressure of the fluid, causing bubbles to form, and then collapse slightly downstream as the flow straightens and pressure returns.

A similar phenomenon occurs at ordinary pipe entrance from tanks. As the water rounds the bend and increases velocity in exiting the tank, the flowlines cause a slight contraction of the stream at pipe entrance. This creates a low pressure area along the inside wall of the pipe where vapor forms and cavitation can occur.

SPECIALIZED PUMPING SYSTEMS
Submersible Pumps

Submersible pumps see increased service each year for many applications. They are now commonplace for sewage and stormwater pumping; submersibles remove urban drainage water to and from retention basins, are used to irrigate farmland and golf courses, and replenish wetlands. They provide good low pressure pumping for transfer applications: pond to pond, from ditches, and for construction dewatering. In potable water service, submersibles provide high pressure for deep well pumping when equipped with multi-stage impellers. Since the most characteristic and practical use for submersible pumps is for wastewater lift station pumping, we should especially consider this.

Reviewing the two basic types of lift stations.

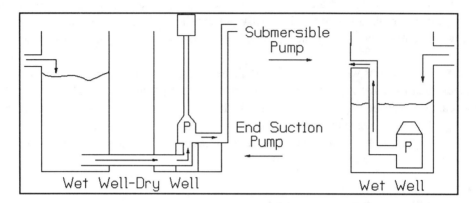

The double well system employs an end suction pump, which takes water with a positive head from the separate wet well, and pumps it through a force main to the surface. The dry well section of the installation houses pump, motor, control panel, and all associated equipment. This structure often extends well above ground as the pumphouse.

The single well system has a wet well only, with submersible pump installed at the

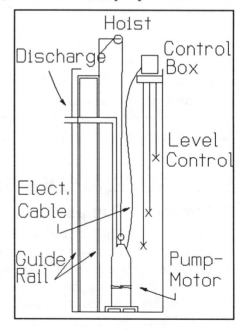

bottom. The pump is submerged, but control panel, including starters and alarm system for high water level, is above ground. Alarm may be visual, audible, or remote readout through telemetry to treatment plant. The pump is installed on guide rails, can be easily hoisted out for service, and is equipped with a quick disconnect coupling for removal: a sealing flange automatically releases and re-engages the discharge piping. The submersible pump lift station can be built in place or prefabricated, and is sized by the inflow during the peak period of the day. The high water level is determined by the elevation of the basement most likely to flood, and the low water level is set by the elevation of the incoming sewer and the top of the pump casing.

Submersible pumps are vertical, close-coupled, heavy duty centrifugal pumps designed to work submerged in the water which they are pumping. There is no suction piping, and the pump impeller is at the very bottom of the unit. Both motor and pump are submerged. The motor is watertight, and double sealed from the surrounding water by oil or air. Better quality motors are equipped to signal an alarm when a problem senses leakage in the outer seal. Wastewater submersibles are designed to be non-clogging, and should easily pass 3 inch solids. A special design is the grinder pump, which incorporates a stainless steel knife blade on the inlet side of the impeller to chop solids to a slurry for pumping through small diameter force mains. Submersible pump capacities range from 50 to 7000 gpm; heads range from 15 to 200 ft., and power requirements range from ½ to over 100 hp.

There are multiple advantages to using a submersible pump:

Space and equipment savings: There is no need for a pumphouse. Control panels at ground level above the wet well can be locked and/or simply sheltered, or they may be installed at a remote location. Ventilation, lights, dehumidifier, sump pump of the pumphouse are not needed. This adds aesthetic value, as well as construction and operating cost savings to the installation.

Submersible pumps are quiet. They operate below ground and underwater. There is no danger of the motor flooding. Guide rail system enables easy servicing. With no suction piping, and the pump on only when water pressure (depth) is sufficient, problems caused by lack of suction head are eliminated. The surrounding water constantly cools both pump and motor, extending life and making them quite efficient.

Pressure Sewer Systems

If the terrain is rocky, or the water table extremely high, or it is necessary to minimize sewer detention time, a gravity flow wastewater collection system may not be practical; a pressurized sewer system may be the best choice. It is easiest to view this as the opposite of a water distribution system. Sewage is pumped directly from each individual house into one or more trunk lines which lead to the treatment plant.

There are two main types of pressure sewer systems in use:

Grinder Pump System - each home is equipped with a wet well (about 50 gallons) in which is a submersible grinder pump. Wastewater flows by gravity out of the home into the wet well, built just below house level. From there it is pumped into the system and to the treatment plant.

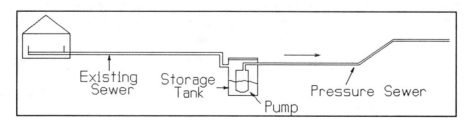

STEP System - (Septic Tank Effluent Pump). This system was originally designed to make use of already existing septic tanks. Water flows from the home to the septic system, and the effluent runs into a small holding tank instead of a drainfield. From there it is pumped by a non-clogging submersible pump into the collection system. This eliminates the need for the grinder pump, with its extra maintenance.

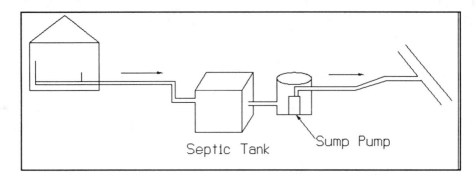

Pressure sewer systems have the advantage of eliminating most of the groundwater infiltration encountered in gravity systems. Because the small diameter PVC piping is buried just below the frost line, installation cost can be lower than a gravity system. Disadvantages are in added operation and maintenance costs incurred due to the need for mechanical equipment at each point of entry to the system.

To estimate flow in sizing pumps for these systems:

Numbers of homes to be serviced x 3.5 people/home x 100 gpd/cap. = gpd to be pumped

Deep Well Pumps

The typical deep well pump is a vertical turbine pump. This is a centrifugal pump resting on a solid foundation with a shaft extending all the way down the well. Multi-stage impellers are arranged near the bottom of the shaft. As with submersible pumps, vertical turbine pumps have no suction piping; the well water level must be over the lowest pump bowl. Each impeller stage increases head as the water moves upward. The well must be straight and vertical, and correct alignment of shaft sections and column pipe is very important.

A well pump follows its characteristic curve like any other centrifugal pump. Capacity is dependent on head, and well drawdown is part of that head. The pump curve applies to each impeller separately. Drawdown should remain constant while pump is in service. If the drawdown is too great air will begin to enter the pump, decreasing the flow rate. As flow decreases, drawdown will start to decrease, allowing the flow to increase - producing a cycling effect of high and low flows. Weather may have a significant effect on well pumping. Dry years lower the water table; pump head increases, flow decreases.

Submersible pumps with multi-stage impellers are also in use as deep well pumps.

Jet Pumps

These are centrifugal pumps which provide extra suction lift for the operation of small wells which must draw water up from greater than 20 ft. below the pump. Installed on pump discharge in small ones, and at well bottom on larger pumps, is a venturi apparatus which increases the velocity of the water enough (extra Velocity Head provided) to obtain an adequate NPSHA.

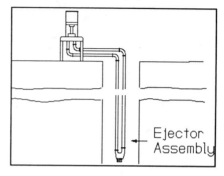

Well Pumps - Plotting Field Efficiency

This refers to the efficiency of a pumping system in operation. It should be checked yearly if the pump is running well.

The pump curve shows the conditions of head and flow that will yield maximum efficiency, at design conditions. Anything off this will be less efficient, but operating conditions should still fall on the curve.

Read the metered gpm. Refer to pump curve for the head. Now measure the head: measure distance to pumping water level, from ground. Add this to the lift above ground. Add the head losses. This is the Total Head. This applies to the entire pump. Divide it by the number of impellers. Now it should match the head read on the curve. If it doesn't, the headloss calculation is incorrect. There may be more headloss than recognized because pump is old, clogged, C factor is off, etc. Doing this helps to identify problems in the system. This is your operating point. Check pump efficiency. To obtain motor efficiency, read the electric meter on the pump motor. Being an accumulator, it will be ticking off Kilowatts. Time the number of Kilowatts per minute. Multiply by 60 to obtain Kw/hr. Divide by .746 to change it to Hp. This is Motor Hp, the amount of electricity actually drawn to run this pump. Now refer to Hp for the operating point of the pump curve (bhp). Multiply it by the pump efficiency (or calculate gpm x hd/3960) to obtain water horsepower. This is work the pump is actually doing.

Compare one against the other, whp, mhp, and you have wire to water efficiency, the FIELD EFFICIENCY.

To obtain motor efficiency, refer back to curve for pump efficiency.
Calculate motor efficiency (Field Efficiency = motor eff. x pump eff).

PUMP APPURTENANCES

Pressure Gage

A pressure gage on a pump should be installed on the discharge piping as far from the pump as practical. If mounted on the pump casing, it will not read accurately, as the energy in the liquid has not yet been fully converted to pressure. A properly operating gage can be a handy tool to diagnose pump problems. If readings begin to show a steady increase in pressure, there may be an obstruction building downstream from the gage. If gage readings show a steady decrease in pressure, check the possibility of a leak or break in the pipe downstream. If this turns out to be the case, turn off the pump and read the gage pressure. The reading will show the elevation of the break.

A discharge gage can be used to check the actual shutoff head. Close the discharge valve, and leave the pump running. The gage will read the true shutoff head. If this value is lower than the shutoff head on the pump curve, there may be internal pump wear or damage.

A discharge gage will give you the approximate flow rate. Read the gage, and refer to pump curve for the flow. (If there is suction head, subtract from gage reading); if there is suction lift, add to gage reading.

Control Valves

These specialized valves provide pressure and flow regulation for protection of pump and system, and are a valuable addition to any pump installation. One of the most important functions of control valves is to assure slow open and close valve operations under normal conditions of pump start/stop, to prevent system shock. Many types of valves are in use for flow control: plug, ball, butterfly, globe and check valves each have applications. Economics and operational requirements will determine which is purchased.

Below is explained a simplified control application, composed of three parts:

Diaphragm Actuator - converts pressure energy from the water to mechanical energy to close the flow valve.

Flow Valve - a globe valve, which provides tight shutoff, and long open/close operation.

Solenoid Valve - a separate small valve, electrically operated, which is set to operate on a timed basis, with pump start/stop, or in response to a limit switch.

All three components are part of the control valve installation. This type is called an Electro-Hydraulic Diaphragm Control Valve, is basically self-contained, and is used on booster pump discharge lines. Its function is to control surge by slowly opening and closing the discharge valve in response to quick start/stop of the pump. The diaphragm is directly connected to the globe stem. The solenoid opens a small vent to atmosphere just above the diaphragm to control function of diaphragm and valve.

OPERATION:

Pump Is Off - there is pressure in the line on upstream side of valve, which is allowed into valve chamber through solenoid apparatus. Diaphragm extends outward; flow valve is closed.

Pump Starts (suddenly) - upon start, solenoid is activated and vent opens; pressure is vented from line. Diaphragm relaxes and flow valve slowly opens.

Pump Stops (suddenly) - upon stopping pump - before it actually stops, the solenoid is deactivated and vent closes; pressure builds up in valve chamber above diaphragm, pushing it outward and slowly closing the flow valve. As closure completes, a limit switch signals pump to turn off, now eliminating the source of pressure. Valve has already closed, slowly.

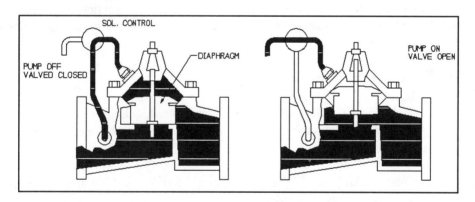

PUMP OPERATION GUIDELINES

In selection of any pump, know the exact nature of the liquid you will be pumping (specific gravity, vapor pressure, entrained air, temperature, corrosive properties, type of concentration of solids). In wastewater applications, solids are most important; at a point they change the specific gravity, and it will require more horsepower to do the same amount of work. For abrasive solids, special pump linings will be needed. Stringy materials may require special impellers.

Advise the manufacturer of the water characteristics, and the capacity needed.

Analyze suction and discharge conditions. Know whether service will be continuous or intermittent, what type of drive will be used, and what power supply is available.

Be aware of location conditions (space, elevation, population density), capital and operating costs, and transportation limitation.

Do you want a constant or variable speed drive?

Make sure spare parts and standby equipment will be available.

Select the pump which will yield the required capacity under the worst operating conditions.

Install the pump in light, dry, clean location, and as low as possible. The foundation should be rigid, the driver properly aligned. Appurtenances should include vent valves at high points on the line, drain valves, relief valves, bypass connections, gages, flowmeter, and source of seal water. Suction piping should be of large diameter and minimum length, and should always be larger than inlet port of pump.

Follow instruction manual on startup and shutdown. Operate the equipment within proper range of flows, pressures, temperatures. Do not throttle suction to reduce pump capacity. Do not use excessive lubricant or cooling water. Do not run it if it is noisy. Avoid sudden temperature changes. Run spare equipment occasionally. Set up scheduled maintenance and inspections. Run the pump at highest efficiency possible to prevent bearing failure.

A pump handles liquids. Keep air out!

Do not open equipment unless diagnosis indicates need. Follow manufacturer's instructions. Clean internal surfaces and repaint where indicated. Replace gaskets. Examine for corrosion, wear, etc.

The weakest part of the pump is the stuffing box. Restore packing areas to proper condition. Be careful mounting bearings.

Keep a complete record of inspection and repairs.

PROBLEMS

1. A pump has an impeller diameter of 14 inches and operates at 1750 RPM.
    A. What is the velocity of the water circulating at the impeller rim?
    B. What is the pressure developed by this pump?

2. A centrifugal pump delivers 300 gpm when its impeller speed is 1750 RPM and the total dynamic head is 57 psi. Determine:
    A. The pump's capacity if speed is reduced to 1250 RPM.
    B. The developed head in the pump if the speed is reduced to 1250 RPM. Assume ideal pump operation.

3. With the pump whose characteristics are represented by curve #2, operating at 1800 gpm, find the change in head, flow, and horsepower when a 6 inch impeller is replaced by a 6½ inch impeller.

4. A industrial process pump is taking 400 gpm from a tank through a 6 inch diameter line from a submerged bottom inlet. The water depth over the inlet is 15 ft; the distance from the tank to the pump is 35 ft; the pump elevation is 8 ft. above inlet. What is the pressure reading at the suction inlet of the pump?

5. NPSHR for a particular pump is 6 ft. Installed into the system described below, determine the NPSHA, and the probability of efficient operation of the pump.
    > Suction lift 15 ft.
    > Suction head loss 12 ft.
    > Flow 300 gpm
    > Suction side inlet 4 inches
    > Vapor pressure 2 ft.

6. A 3000 gpm 12 inch diameter turbine well pump is located 4000 ft. above sea level, and is pumping water at a maximum temperature of 90 degrees F. Suction head losses are 2 ft; water level is never less than 8 ft. above the first stage impeller. What is the NPSHA?

7. NPSHR is 7 ft. Flow is 800 gpm through 200 ft. of 8 inch diameter suction pipe. The lift is 15 ft. on the suction side. The pump is pumping cold water at sea level.
    A. What is the NPSHA?
    B. Is it adequate to lift the water without cavitation?

8. A new pumping installation is being designed: Water at 110 degrees F is to be pumped at a rate of 55 gpm from a holding tank which is open to atmosphere. The gage pressure at the end of the discharge line must be at least 56 psig. The discharge end is 30 ft., and the pump suction is 6 ft. above the water level in the holding tank. The discharge line will consist of 450 ft. of 2 inch diameter new steel pipe. Suction pipe is 2 inch diameter and there are four 90 degree standard elbows in the line. The pump efficiency is 63%. Determine the following:
    A. Head which the pump must develop.
    B. Horsepower input
    C. NPSHA

9. The center of the intake end of a suction pipe is 5 ft. below the water surface in the river. The pipe has a uniform rise of 1 ft. per 100 ft. to the pump. Friction loss at pipe entrance is 5 ft; friction loss in the pipe is 1 ft. per 1000 ft. What is the greatest length the pipe can have without causing the pressure to drop more than 6 psi below atmospheric pressure?

10. Referring to pump shown on curve #2 (Chapter 13)
    Flow = 1600 gpm
    Suction Head 25 ft.
    Discharge reservoir elevation 100 ft.
    Discharge line 2500 ft. of 10 inch diameter 10 yr. old CIP
    Timed .85 Kw./min. on electric meter
    What is the field efficiency?

SOLUTIONS

1. **Answer**   A. 107 ft./sec.
                   B. 178 ft.

   A. Change diameter to ft.; find distance traveled for each revolution of impeller; solve for distance traveled/sec.

      14 inches = 1.17 ft.

      C = 3.14 x diam.

      C = 3.14 x 1.17

      C = 3.67 ft. (distance a point at impeller tip travels per revolution.)

      Velocity = rev./min. x ft./rev.

      V = 1750 x 3.67

      V = 6423 ft./min.

      V = 107 ft./sec.

   B. All the velocity head is changed to pressure head upon exiting pump.

      $$Vh = \frac{V^2}{64.4}$$

      $$Vh = \frac{197^2}{64.4} = 178 \text{ ft. of pressure created by pump}$$

2. **Answer**   A. 214 gpm
                 B. 67.3 ft.

   A. Capacity:

   $$\frac{\text{New RPM}}{\text{Old RPM}} \times Q = \text{New Q}$$

   $$\frac{1250}{1750} \times 300 = \text{New Q}$$

   $$214 \text{ gpm} = \text{New Q}$$

   B. Head: (57 psi = 132 ft.)

   $$\left(\frac{\text{New RPM}}{\text{Old RPM}}\right)^2 \times \text{head} = \text{New Head}$$

   $$\left(\frac{1250}{1750}\right)^2 \times 132 = \text{New Head}$$

   $$67.3 \text{ ft.} = \text{New Head}$$

---

3. **Answer**   150 gpm, 19 ft. head, 17 hp change

   Refer to curve for head and hp with current impeller; then calculate change.

   $$Q = 1800 \text{ gpm}$$

   $$TDH = 110 \text{ ft.}$$

   $$Bhp = 62 \text{ hp}$$

   $$\frac{\text{New diam.}}{\text{Old diam.}} \times Q = \text{New Q}$$

   $$\frac{6.5}{6} \times 1800 = \text{New Q}$$

   $$1950 \text{ gpm} = \text{New Q}$$

   Pump delivers 150 more gallons/minute with larger impeller.

   $$\left(\frac{\text{New diam.}}{\text{Old diam.}}\right)^2 \times \text{head} = \text{New Head}$$

   $$\left(\frac{6.5}{6}\right)^2 \times 110 = \text{New Head}$$

   $$129 \text{ ft.} = \text{New Head}$$

   Pump lifts water 19 ft. higher with larger impeller.

$$\left(\frac{\text{New diam.}}{\text{Old diam.}}\right)^3 \text{ x Bhp } = \text{ New Bhp}$$

$$\left(\frac{6.5}{6}\right)^3 \text{ x } 62 = \text{ New Bhp}$$

$$79 \text{ hp } = \text{ New Bhp}$$

Pump uses 17 hp more with larger impeller.

---

4. **Answer**   2.7 psi

Draw diagram. Suction gage reading equals static suction head minus head losses.

Solve for HL:

400 gpm = .89 cfs

Assume C = 100

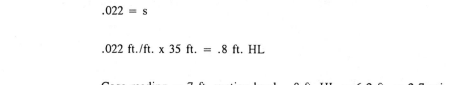

$$Q = .435 \quad C \quad d^{2.63} \quad s^{.54}$$

$$.89 = .435 \quad 100 \quad .5^{2.63} \quad s^{.54}$$

$$.022 = s$$

.022 ft./ft. x 35 ft. = .8 ft. HL

Gage reading = 7 ft. suction head - .8 ft. HL = 6.2 ft. = 2.7 psi

---

5. **Answer**   6 ft. Poor operation; cavitation likely

Solve for Velocity Head.

$$Q = 300 \text{ gpm } = .67 \text{ cfs}$$

$$Q = A \quad V$$

$$.67 = .785 \text{ x } .33^2 \quad V$$

$$7.8 \text{ ft./sec. } = V$$

$$Vh = \frac{V^2}{2g} = \frac{7.8^2}{64.4} = .95$$

NPSHA = (Atmospheric Press. = Vh) - (Lift + Losses + vapor pressure)

NPSHA = (34 + .95) - (15 + 12 + 2)

NPSHA = 6 ft.

NPSHA = NPSHR

---

6. **Answer**   35 ft.

All values are recorded in feet of water Absolute.

We are only considering the suction side of the pump; pump is immersed to 8 ft. in the water (8 ft. suction head); there is no suction lift; suction head must be entered as a negative value.

Vapor pressure: referring to chart, VP at 90° F = 1.6 ft.

Atmospheric Pressure: referring to chart, at 4000 ft. = 29.4 ft.

Velocity Head: solve for velocity (3000 gpm = 6.7 cfs)

$$Q = A \quad V$$

$$6.7 = .785 \times 1^2 \quad V$$

$$8.5 \text{ ft./sec.} = V$$

$$Vh = \frac{V^2}{2g} = \frac{8.5^2}{64.4} = 1.1 \text{ ft.}$$

NPSHA = (Atmos. Press. + Vh) - (Lift + Losses + vapor press.)

NPSHA = (29.4 + 1.1) - (-8 + 2 + 1.6)

NPSHA = 35 ft.

The pump should operate efficiently as long as NPSHR is less than 35 ft.

---

7. **Answer**   A.   15 ft.
                   B.   yes

A. Lift = 15 ft.; Atmos. Press. = 34 ft.; Vapor Pressure = .5 ft.

Solve for Vh, then losses, then NPSHA (800 gpm = 1.8 cfs).

$$Q = A \quad V$$

$$5.1 = .785 \times .67^2 \quad V$$

$$1.8 \text{ ft./sec.} = V$$

$$Vh = \frac{V^2}{2g} = \frac{5.1^2}{64.4} = .4 \text{ ft.}$$

Assume C=100

$$Q = .435 \quad C \quad d^{2.63} \quad s^{.54}$$

$$1.8 = .435 \quad 100 \quad .67^{2.63} \quad s^{.54}$$

$$.02 = s$$

.02 ft./ft. x 200 ft. = 4 ft. HL in suction pipe

NPSHA = (Atmos. Press. + Vh) - (Lift + Losses + vapor press.)

NPSHA = (34 + .4) - (15 + 4 + .5)

NPSHA = 15 ft.

B. Since NPSHA (15 ft.) is greater than NPSHR (7 ft.), the pump is adequate to lift the water without cavitation.

NOTE: That when pumping cold water at sea level, the lift and losses are the major factors which decrease the available pressure.

---

8. **Answer**   A. 222.6 ft.
                 B. 4.9 hp
                 C. 24.9 ft.

Draw diagram.

A. TDH = Lift + Losses

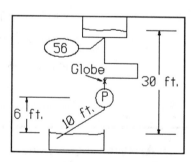

Lift:

30 ft. + pressure at discharge end (56 psi)

56 psi = 129.4 ft.

30 ft. + 129.4 ft. = 159.4 ft.

Losses:

New steel pipe C=140

55 gpm = .123 cfs

$$Q = .435 \quad C \quad d^{2.63} \quad s^{.54}$$

$$.123 = .435 \quad 140 \quad .167^{2.63} \quad s^{.54}$$

$$.062 = s$$

Minor Loss: 450 ft. discharge pipe + 10 ft. suction pipe

straight pipe	460 ft.
1 open globe	50 ft.
4 elbows	20 ft.
1 borda entrance	6 ft.
	536 ft. of total pipe

.062 ft./ft. x 536 ft. pipe = 33.2 ft. HL in total line

TDH = 30 ft. lift + 33.2 ft. losses + 159.4 ft. pressure = 222.6 ft. TDH

B.   Solve for Bhp.

$$\frac{gpm \times head}{3960} \times whp$$

$$\frac{55 \times 222.6}{3960} \times whp$$

$$3.1 = whp$$

$$\frac{3.1 \; whp}{.63} = 4.9 \; Bhp$$

C.   Vapor pressure at 110° F (from chart) = 3 ft.

Solve for Vh:

$$Q = A \quad V$$

$$.123 = .785 \times .167^2 \quad V$$

$$5.6 \; ft./sec. = V$$

$$Vh = \frac{V^2}{2g} = \frac{5.6^2}{64.4} = .5 \; ft.$$

Solve for suction losses:

.062 ft./ft. (slope) x 10 ft. (length suction) = .62 ft. HL in suction

NPSHA = (Atmos. Press. + Vh) - (Lift + Losses + Vapor Press.)

NPSHA = (34 + .5) - (6 + .62 + 3)

NPSHA = 24.9 ft.

9. **Answer**   1270 ft.

Draw diagram -- Maximum allowable drop in pressure is 5 ft. for suction head plus 6 psi (14 ft.) below atmospheric pressure. (19 ft. drop allowed). Lift and losses must not exceed this.

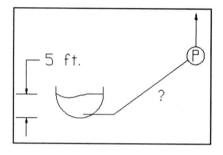

Lift: 1 ft. for every 100 ft.

Losses: 5 ft. for fitting + 1 ft. for every 1000 ft. of length.

First, subtract fitting loss from maximum allowable pressure drop. This loss will be present no matter how long the pipe is.

19 ft. - 5 ft. = 14 ft.

The lift plus friction losses must not exceed 14 ft.

First try:

If 1400 ft. of pipe will lift the water 14 ft., then this length alone, with no friction losses, will drop the pressure 14 ft.

Friction losses in this length of pipe would be 1.4 ft. (1 ft./1000 ft.). That gives us 15.4 ft. of pressure loss. No good.

Second try:

1250 ft. of pipe would lift the water 12.5 ft. This drops the pressure 12.5 ft. Friction loss in 1250 ft. of pipe is 1.25 ft. This gives us 13.75 ft. pressure loss. We can do better.

Third try:

1270 ft. of pipe lifts the water 12.7 ft., dropping pressure 12.7 ft. Friction loss in 1270 ft. pipe is 1.27 ft. This yields a pressure drop of 13.97 ft. Close enough.

---

10. **Answer**   73%

Draw diagram. With metered flow, refer to curve for TDH. Calculate TDH; should be the same as on curve. Refer to curve for Bhp & pump efficiency at that flow. Calculate Mhp; calculate motor efficiency, then wire to water efficiency.

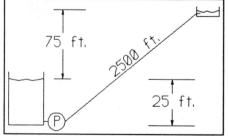

Flow = 1600 gpm (3.6 cfs)

From curve, TDH = 125 ft.

Bhp = 59 hp

pump efficiency = 85%

Calculate TDH:

$Q = .435 \quad C \quad d^{2.63} \quad s^{.54}$

$3.6 = .435 \quad 110 \quad .83^{2.63} \quad s^{.54}$

$.02 = s$

.02 ft./ft. x 2500 ft. = 50 ft. HL due to friction

TDH = Lift + Losses

TDH = 75 + 50

TDH = 125 ft.

OK. This checks out.

Calculate Mhp:

.85 Kw/min. x 60 min./hr. = 51 Kw/hr.

$$\frac{51 \text{ Kw/hr.}}{.746 \text{ Kw./hp}} = 68.4 \text{ Mhp}$$

Motor efficiency $= \dfrac{\text{Bhp}}{\text{Mhp}} = \dfrac{59}{68.4} = 86\%$ efficient

Motor eff. x pump eff. = wire to water efficiency

.86 x .85 = .73

Field efficiency = 73%

CHAPTER 15

POSITIVE DISPLACEMENT PUMPS

Whereas centrifugal pumps are meant for low pressure, high flow applications, positive displacement pumps can achieve greater pressures, but are slower moving, low flow pumps. Positive displacement pumps are used for pumping sludges and slurries and other high density or viscous fluids for which other types of pumps are inappropriate. Tolerances are close between stationary and moving parts, and flow is constant. A given volume of liquid is positively displaced with each pump cycle; there is no backwards slippage of fluid, and high shutoff pressures are developed. When pumping head increases from a blockage, or when system head increases, the pump will still move the same amount of fluid, but will draw more horsepower, and create high discharge pressure in overcoming that head.

This is quite different from the operation of the centrifugal pump, which delivers less flow and uses less horsepower under high head conditions. With a positive displacement pump, total blockage of pump or line, whether caused by foreign material or a closed valve, will cause failure of pump, line or motor. Never start a positive displacement pump with the discharge valve closed!

These pumps operate at much slower speeds than centrifugal pumps. They are much larger, heavier, and more costly. The large number of moving parts (valves, cams, reciprocating components) make them more complex, and the close running clearances make them more subject to wear, with resulting parts replacements necessary. Efficiency is lower and maintenance cost higher than in the centrifugal pump.

There are two basic categories of positive displacement pump: the reciprocating pumps, and the rotary pumps. Under reciprocating pumps are generally included the plunger, or piston pump, the diaphragm pump, and the peristaltic pump. Under rotary pumps are included the rotary gear pump, the vane pump, the progressive cavity pump, and the Archimedes screw.

RECIPROCATING PUMPS

These pumps have suction and discharge check valves, and the flow is cyclical; there is a to and fro motion; flow pulsates.

Plunger Pumps

This is a reciprocating pump that uses a piston to force liquid from the suction side to the discharge side of the pump. It is most often seen in wastewater applications to move primary sludge. The piston is located in the middle of the length of horizontal large diameter pipe. There are check valves on either side. As the piston rises, liquid is sucked into the pump. The suction check valve opens to allow liquid to enter the piston chamber, and the discharge check closes to prevent previously discharged liquid

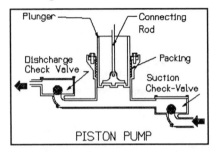

from backing up into the pump. As the piston lowers, the suction check closes and the discharge check opens, allowing forward movement of the liquid. Clearance between the piston and chamber walls is filled by lubricated packing, in a packing gland. The piston is connected to the driver by a connecting rod, and there are pressure gages on suction and discharge side of the pump. Checking these will give a good indication of what is going on inside the pump. Pressure relief devices are usually associated with this type of pump, for use when blockages occur and pressure builds up; often they are connected to an alarm. On sludge pumps, the check valves are balls made of hard material; both valve seat and balls will wear and are replaceable. The presence of the check valves make this type of pump essentially self-priming.

Another type of piston pump is referred to as the lift pump. This is the typical backyard hand operated pump. A four stroke cycle is necessary to operate it. A bucket valve is built into the piston and moves up and down with it.

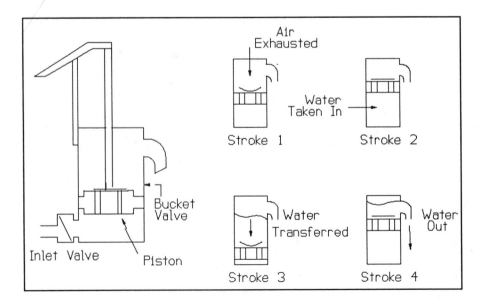

A. Air exhaust - piston descends to bottom of cylinder, forcing out air.
B. Water inlet - upward stroke - vacuum is created. Water flows into cylinder.

C. Water transfer - downward stroke - water flows through the bucket valve, and is transferred to upper side of piston.

D. Water discharge - piston rises - water is discharged from pump.

A piston pump which is a deep water well pump has the piston at the bottom of the well. Like centrifugal pumps, positive displacement pumps with a suction lift generally are restricted to about twenty feet of practical lift.

Many large pumping stations equipped with piston pumps were built during the nineteenth century. Enormous engines were employed. The development of the electric motor and centrifugal pump provided a compact pumping unit which has virtually eliminated these pumps with their pulsating flow from water service.

Other types of piston pumps are used in industrial applications, often with spring actuated multiple pistons which move rapidly. These are in use for their ability to develop very high pressures, and are not meant for solids handling.

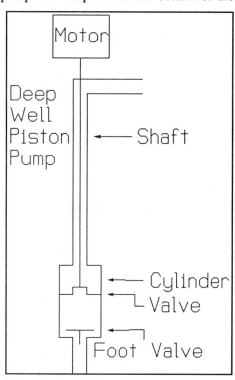

Diaphragm Pumps

In water utilities, diaphragm pumps are most often used for chemical feed. The pumping chamber is in a vertical position with the check valves above and below the diaphragm chamber. The principle of operation is the same as that of the plunger pump. The pumping mechanism is controlled by a rod which moves back and forth on a cam, and pushes the diaphragm in and out, moving the liquid. Percent stroke of the diaphragm is controllable by setting, and is the means of controlling flow. Checks are sometimes shaped as discs, rather then balls, and are much lighter weight than those on plunger sludge pumps. If the pump is located above the liquid to be pumped, priming may be necessary. Diaphragms and checks will wear easily, and are replaceable. There

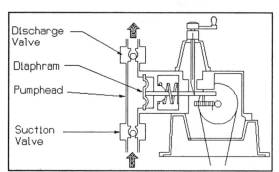

are several variations of the diaphragm pump available and in use, referred to as chemical metering pumps.

Peristaltic Pumps

Peristaltic pumps are not true reciprocating pumps, as they do not have a back and forth action, but they do however, produce a pulsating flow, and so are included here. These are mainly used as chemical feed pumps also, usually for very small flows, and are controlled by a set of fingers, or a moving ball, which repeatedly squeezes flat a length of tubing filled with the liquid to be pumped, expelling it. As the tubing expands again, it fills with incoming liquid, which waits to be squeezed forward again. There are no valves associated with this type of pump. Movement is strictly by peristaltic action. Tubing is replaced as necessary.

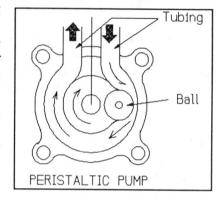

PERISTALTIC PUMP

ROTARY PUMPS

Rotary pumps are positive displacement pumps which are primarily used as a source of fluid power in hydraulic systems, for auto, aircraft, and mobile equipment applications. A few types are in common use in water utilities.

Progressive Cavity Pumps

Chemical slurries, polymers, and sludges are often moved with progressive cavity pumps, a type of rotary pump which can achieve pressures up to 2000 psi, depending on pump length. The liquid travels in the spaces between a rigid rotor (hardened chrome steel) and a flexible stator (rubber). It operates in a manner something like a snake swallowing a frog; as each section closes, it opens the section in front of it for the liquid to move into -a cavity progression. The rotor revolves rapidly, capacity is directly proportional to rotor speed and pump size, and a non pulsing flow is produced. There are no valves associated with the pumping, but packing is needed to prevent liquid from leaking out of the pump and up the shaft. Rotor and stator are always in contact as they move, and the liquid pumped does the lubricating. This type of pump should never by run dry, as it will quickly burn up the stator and fail.

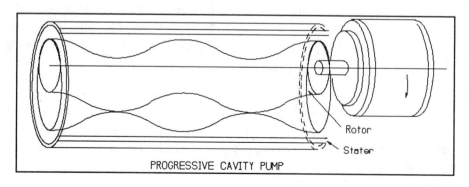

PROGRESSIVE CAVITY PUMP

Rotary Impeller Pumps

Used for oil transfer and hydraulic services, the "impeller" is sometimes soft and flexible, and scoops the liquid in and out of the pumping chamber, rather than imparting velocity. Its appearance is that of a centrifugal pump, but the impeller speed is lower, and the close tolerance between impeller and casing make it positive displacement. Two common types of rotary impeller pumps are:

Rotary Gear Pumps - intermeshing gears, or lobes, trap the fluid between them

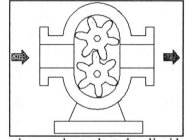

as they rotate. Gear pumps can withstand rugged operating conditions, and are simple in construction, and economical in cost and maintenance. They can handle many types of liquids (including air/liquid mixtures) over a wide range of capacities and pressures, but are best adapted to low pressures and flows under 500 gpm. These are sometimes called Hydraulic pumps, and work well with suction lifts. Because of the constant close meshing of the gear lobes, they are best adapted to liquids having lubricating qualities.

Rotary Vane Pumps - used for high pressure industrial applications, often with

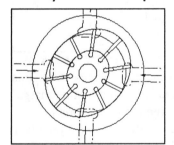

hot liquids. The vanes slide in and out as the pump chamber fills and empties. Vane pumps emphasize quiet operation, long life, and little load on the bearings.

Screw Pumps

The Archimedes Screw was invented by the ancient Greek himself about 287 BC (perhaps the world's first great physicist). This is a large piece of lifting apparatus, not truly a pump at all, but in fact a screw with deeply cut threads, laid at an angle up to 45 degrees, with its lower end submerged in the water. The liquid submergence at entrance, and the speed of revolution control

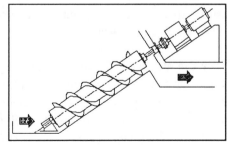

the output. As the pump revolves, the water "crawls" up the screw. Maximum lift is about 25 ft., but the pump can be built in stages for higher lifting. An angle of more than 45 degrees of incline results in backwards slippage of the water, and is inefficient. The screw pump is relatively maintenance free, passes large solids, and is economical to operate. Limitations are with maximum lift, and with size. This type of pump is often seen as a raw wastewater pump at headworks of small treatment facilities. Laid horizontally, screw conveyors are in use for moving grit, dewatered sludges, and other high solids materials.

MISCELLANEOUS PUMPING APPARATUS
Air Pumps

Pneumatic Ejector - The pneumatic ejector is based on reciprocating operation, and is a fill and draw affair. The lift it achieves is based on air pressure, not water pressure. In use in small wastewater lift stations, this type of pumping apparatus is excellent for water containing solids. There are no rotating parts to clog, and it can handle flows up to 300 gpm.

The water flows by gravity into a small vented chamber. When full, the vent closes, compressed air is introduced, and the water is forced out into the discharge line. The vent opens, and new water comes in. The cycle lasts about a minute, and the only machinery is an air compressor.

Air Lift Pump - The air lift pump, also known as the "bubble pump", is also used in small wastewater lift stations, and for pumping activated sludge. Its advantage over the pneumatic ejector is that a smooth constant flow is produced. The chamber includes a vertically positioned discharge pipe which extends to the bottom of the tank, and into the bottom of which compressed air is fed. The bubbles rise and saturate the water in this pipe, decreasing its density. 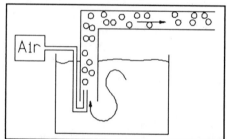 The water filling the tank is heavier, and presses upon the water in the discharge pipe, causing it to move up and out. As new water fills in behind it, it is filled with air, and moves up, thus producing a constant discharge. As with the pneumatic ejector, the only power needed is to run the air compressor, and there are no moving parts. However, lift depends on the differential in hydrostatic pressure, and is limited. Best operation is achieved with horizontal flow.

Vacuum Pumps - Vacuum is any pressure less than atmospheric. Vacuum pumps can be rotary or piston pumps, completely oil sealed, which just pump air out of a chamber, creating vacuum.

They can also be designed as a centrifugal pump, called a Liquid Jet Vacuum Pump. These have a special attachment called an ejector, or eductor. This is a venturi constriction with a side outlet; as velocity increases passing through the venturi, the side outlet which is open to atmosphere, will draw air in, creating a constant useful vacuum as long as the pump operates. The air/water mixture is then discharged into a tank where air and water separate, and only water is recycled through the pump. In water treatment facilities, the eductor is used to move slurries (lime feed, filter sand), to provide gas flow (injector of vacuum chlorinator), and to create vacuum in the laboratory (at sink faucets).

Blowers - Blowers pump air. Used for wastewater aeration service, they are large in size, and power demanding. Design may be positive displacement (rotary gear) or centrifugal (turbine). The centrifugal types run at higher speeds, but are less efficient than the rotary blowers. Drawing in ambient air through a filter, they transport it at low pressures to the underwater point of application through a large diameter discharge pipe called a Header. Near the end of the discharge line, the header splits up into multiple diffusers for application of air to the water. All blowers are noisy.

Hydraulic Ram

The ram lifts water by utilizing the impulse developed when a moving mass of water is suddenly stopped. In other words, it makes use of the pressure created by water hammer. Not an efficient pumping mechanism, a relatively large amount of water must be available at moderate head in order to move a small amount of water to higher head. Water flows from a reservoir to a valve box below, and on out of the box through a waste valve. If that valve is suddenly closed, water hammer pressures develop, which open another valve into a

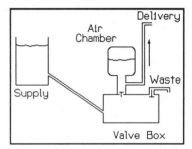

vertical delivery pipe, and some water passes up the pipe. As negative pressure waves return, the delivery pipe valves closes, and water continues on through the box. What keeps it going is the water flow through the waste valve. When well established, it provides enough force to shut the waste valve, starting a new cycle. The hydraulic ram is not really a pump - just a set of valves.

Turbine Generator - The Backwards Pump

A motor converts electrical energy to mechanical energy to turn a shaft. This, connected to a pump shaft, will rotate the pump impeller. The system expends electrical energy to produce water energy (pressure).

The opposite principle, that of the turbine generator, uses water energy to turn a turbine (impeller) connected to a motor shaft, to drive a motor, thus creating electrical energy. In the standard mode, the essential water energy would be provided by a waterfall - on the turbine, to keep it moving. But water pressure is wasted in many places, and can be captured for this type of use. For instance, a large pressure reducing valve between a water supply and a city's water system, reduces and controls pressure so that it remains at a useable level. The excess is wasted by the reducing valve. If a pump were installed there (a turbine with a motor), adapted to run backwards, that excess pressure could be captured and converted to electricity. The water system would still benefit from the operation of the pressure reducing valve, and the electricity created by the adaptation could be used to run other pumps, etc. An ordinary pump used this way would need modifications; too much flow would run the impeller too fast for efficient or safe operation. There is a maximum allowable speed for turbine operation and transmission of power to the motor. Beyond this (runaway speed), damage may result to mechanical parts, and control valves are needed. The backwards pump principle is now being put to use in installations across the country.

Hydropneumatic Storage

When stored water is desired, but elevated storage is not practical, hydropneumatic storage may be the best choice for small community water systems, private wells, industries, apartment buildings - anyplace where capital funds and/or space is limited. In this type of installation, air pressure is in use to equalize water pressure, prevent surges, and minimize pump operation time.

A relatively small tank is employed, which has an air chamber, and a water chamber,

with a bladder or diaphragm separating the two. Pressure sensors are set to a range of pressures at the tank. When pressure in the system is low, the pump operates, filling the water chamber. The diaphragm flexes toward the air chamber, compressing the fixed amount of air in that section. This increases air pressure - to the maximum setpoint, and the pump switches off. As the water is used, the tank empties, the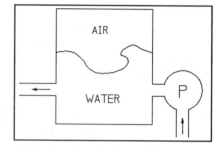

diaphragm relaxes, and pressure decreases to the low setpoint; then the pump switches on again. This allows system pressure to vary over a range of values, decreases operating time for the pump, and the accompanying pressure transients which are associated with a constant on/off mode.

The result is the same as filling and emptying an elevated tank, but air pressure is replacing the water pressure that elevation would provide. Capital cost for hydropneumatic storage is less than elevated storage, but operating costs are greater. A combination of the two modes can be provided by installing the hydropneumatic storage tank on the roof of the building.

PROBLEMS

1. A piston pump chamber is 12 inches in diameter and 18 inches deep. How many gallons of sludge are moved with each stroke?

2. A pump chamber is 10 inches in diameter and 12 inches deep. Pump operates at 1 stroke per second. What is the gpd pumping rate?

3. Primary sludge is pumped to the digester 4 times per day, for a 15 minute pumping cycle each time. If the piston chamber is 12 inches in diameter and 16 inches in depth, pumps at 1 stroke per second, and is set at 3/4 stroke, what is the gpd pumping rate?

4. Primary sludge is pumped to an empty digester, which is 50 ft. in diameter and 30 ft. deep. The pump cylinder is 6 inches in diameter, 8 inches in depth, is set at full stroke, and pumps 1 stroke every two seconds. Sludge is pumped 4 times per day, for a 10 minute pumping cycle each time. At the end of a 2 month pumping period, how deep is the sludge in the digester?

5. A plunger pump with a bore of 5 and 3/16 inches, and a stroke of 2 and 5/32 inches is operating at a rate of 65 revolutions per minute.
   A. What is the gpm delivery?
   B. How many gallons of liquid would be pumped in a 30 day month if the pump operates for 4 hours every day?

6. A piston pump with a 6 inch diameter piston and a 1 ft. depth chamber is set on ½ stroke, at a pumping rate of 500 strokes per minute. It pumps water out of a cylindrical tank 40 ft. in diameter and 50 ft. depth.
   A. What is the drawdown in 4 hours?
   B. If water were also flowing into this tank at 1 cfs, while the pump is running - is the tank filling or emptying, and by how much?

7. Given two piston pumps. Each piston is 8 inches in diameter, and has a stroke of 8 inches. The rate is 50 stroke per minute. They are pumping liquid out of a circular tank 24 ft. in diameter, with a 50 ft. liquid depth. The drawdown after 45 minutes of pumping is 2 ft.
   A. What percent of their capacity are the pumps pumping?
   B. If pumping less than full capacity, what mechanical problem would this most likely indicate?

SOLUTIONS

1. **Answer**   8.81 gallons

   Volume of chamber is moved with each stroke; solve for volume.

   $$\text{Vol.} = .785 \quad d^2 \quad h$$

   $$\text{Vol.} = .785 \quad 1^2 \quad 1.5$$

   $$\text{Vol.} = 1.178 \text{ cu.ft.}$$

   $$\text{Vol.} = 8.81 \text{ gal.}$$

2. **Answer**   349633 gpd

   Solve for volume; change to pumpage.

   $$\text{Vol.} = .785 \quad d^2 \quad h$$

   $$\text{Vol.} = .785 \quad .83^2 \quad 1$$

   $$\text{Vol.} = .541 \text{ cu.ft.}$$

   .541 cu.ft. x 1 stroke/sec. x 60 sec./min. x 1440 min./day = 46742.4 cu.ft./day = 349633 gpd

3. **Answer**   21086 gpd

   Find vol. chamber; mult. by % stroke; change to pumpage.

   $$\text{Vol.} = .785 \quad d^2 \quad h$$

   $$\text{Vol.} = .785 \quad 1^2 \quad 1.33$$

   $$\text{Vol.} = 1.044 \text{ cu.ft.}$$

   $$\text{Vol.} = 7.81 \text{ gal.}$$

   7.81 gal. x .75 = 5.86 gal.

   5.86 gal./str. x 1 str./sec. x 60 sec./min. x 15 min./cycle x 4 cycles/day = 21086 gal./day

4. **Answer**   4.8 ft.

Solve for vol. chamber; get pumpage/2 mo.; solve for liquid vol. in dig.

$$\text{Vol.} = .785 \quad d^2 \quad h$$

$$\text{Vol.} = .785 \quad .5^2 \times .67$$

$$\text{Vol.} = .1315 \text{ cu.ft.}$$

.1315 cu.ft./str. x .5 str./sec. x 60 sec./min. x 10 min./cycle x 6 cycles/day x 60 days = 9468 cu.ft.

$$\text{Vol.} = .785 \quad d^2 \quad h$$

$$9468 = .785 \quad 50^2 \quad h$$

$$4.8 \text{ ft.} = h$$

---

5. **Answer**   A. 12.8 gpm
                 B. 92,312 gal./mo.

Solve for vol. chamber; mult. by RPM.

A. $\text{Vol.} = .785 \quad d^2 \quad h$

    $\text{Vol.} = .785 \quad .432^2 \quad .18$

    $\text{Vol.} = .0264 \text{ cu.ft.}$

    $\text{Vol.} = .197 \text{ gal.}$

    .197 gal./str. x 65 str./min. = 12.8 gpm

B. 12.8 gpm x 60 min./hr. x 4 hr./day x 30 day./mo. = 92,312 gal./mo.

---

6. **Answer**   A. 9.4 ft.
                 B. Filling - by .18 cfs

A. Solve for volume chamber; get pumpage/4 hr.; use vol. tank to solve for h.

    $\text{Vol.} = .785 \quad d^2 \quad h$

    $\text{Vol.} = .785 \quad .5^2 \quad .5$

    $\text{Vol.} = .098 \text{ cu.ft.}$

.098 cu.ft./str. x 500 str./min. x 60 min./hr. x 4 hr. = 11775 cu.ft./4 hr.

$$\text{Vol.} = .785 \quad d^2 \quad h$$

$$11775 = .785 \quad 40^2 \quad h$$

$$9.4 \text{ ft.} = h = \text{drawdown}$$

B. Solve for flow (cu.ft. pumped/sec.) out; compare against flow in.

$$\frac{500 \text{ str./min.}}{60 \text{ sec./min.}} = 8.33 \text{ str./sec.}$$

.098 cu.ft./str. x 8.33 str./sec. = .82 cu.ft./sec.

1 cfs in. - .82 cfs out = .18 cfs in

Tank is filling - by .18 cfs.

---

7. **Answer**   A. 85% capacity
   B. checks slipping; piston worn; set at full stroke?

A. Solve for pumped volume at 100% capacity; solve for volume actually pumped (drawdown); compare.

$$\text{Vol.} = .785 \quad d^2 \quad h$$

$$\text{Vol.} = .785 \quad .67^2 \quad .67$$

$$\text{Vol.} = .236 \text{ cu.ft.}$$

.236 cu.ft./stroke x 50 strokes/min. x 7.48 gal./cu.ft. x 45 min. x 2 pumps
= 7947 gallons pumped theoretical

$$\text{Vol.} = .785 \quad d^2 \quad h$$

$$\text{Vol.} = .785 \quad 24^2 \quad 2$$

$$\text{Vol.} = 904.3 \text{ cu.ft.}$$

$$\text{Vol.} = 6764.3 \text{ gallons drawdown}$$

6764.3 gal. drawdown/7947 theoretical pumpage = 85% capacity

APPENDIX I

CONVERSIONS

Weight to Volume: 1 gram = 1 ml. (water only)

8.34 lb. = 1 gallon (water only)

62.4 lb. = 1 cubic foot (water only)

Volume: 7.48 gal. = 1 cu.ft.

1000 liters = 1 cubic meter

3.785 liters = 1 gallon

Linear: 5280 ft. = 1 mile

Time: 1440 minutes = 1 day

Flow: 1 MGD = 1.55 cfs

Pressure: 1 psi = 2.31 ft. of water depth

1 ft. of water depth = .433 psi

Power: 1 hp = .746 Kw

APPENDIX II

IMPORTANT FORMULAS

Area of a circle: $A = .785 \ d^2$

Perimeter of a circle: $P = 3.14 \ d$

Volume of a cylinder: $Vol. = .785 \ d^2 \ h$

Surface Area of Cylinder Wall: $A = 3.14 \ d \ h$

Pressure: $P = w \ h$

Force: $F = P \ A$

Equation of continuity: $Q = A \ V$

Bernoulli's Formula: $Ph_1 + Vh_1 + Z_1 = Ph_2 + Vh_2 + Z_2 + HL$

Velocity Head: $Vh = \dfrac{V^2}{2g}$

Horsepower: $\dfrac{gpm \ x \ hd}{3960} = whp$

Hazen-Williams Formula: $Q = .435 \ C \ d^{2.63} \ s^{.54}$

Slope: $s = \dfrac{HL}{L}$

Manning's Formula: $V = \dfrac{1.486}{n} \ R^{.66} \ s^{.5}$

Hydraulic Radius: $R = \dfrac{wetted \ area}{wetted \ perimeter}$

Hydraulic Radius (full pipe): $R = \dfrac{d}{4}$

Flow through orifice: $Q = C_d \ A \ \sqrt{(Ph_1 - Ph_2) \ x \ 2g}$

Flow through venturi: $Q = C_d \ A \ \sqrt{\dfrac{(Ph_1 - Ph_2) \ x \ 2g}{1 - \left(\dfrac{d_2}{d_1}\right)^4}}$

Weir Flow (rectangular): $\quad Q = 3.33 \quad L \quad h^{1.5}$

Weir flow (90° V-notch): $\quad Q = 2.5 \quad h^{2.5}$

Flow through Parshall Flume: $\quad Q = 4 \quad W \quad H_n^{1.52} \quad W^{.026}$

Trajectory of a jet: $\quad Q = \dfrac{4\ A\ X}{\sqrt{y}}$

Available Net Pos. Suction Head: $\quad$ NPSHA =
(Atmos. Press. + Vh) - (Lift + Loss + vp)

APPENDIX III

PUMP TROUBLESHOOTING

No Water Delivered

pump not primed - or incompletely primed
discharge head too high - running over shutoff head - check friction losses
suction lift too high - check gage and size of suction pipe
inlet of suction pipe insufficiently submerged
air in through stuffing box or suction piping
plugged pump or piping
rotation or impeller backwards

Not Enough Water Delivered

insufficient prime
partial plugging
speed too slow
discharge or suction lift too high
air in through packing or suction piping (check packing with match; replace packing;
    check position of lantern ring; water seal piping may be plugged)
foot valve too small or plugged - check strainer
entrained air coming in through suction tank
drawdown too great (well pump)
defective or worn impeller or wear rings

Discharge Pressure Too Low

entrained air
speed too low
impeller diameter too small
impeller clearance too great
discharge gage in the wrong place
mechanical defects - deformed casing

Discharge Pressure Too High

too much friction loss - are valves wide open?

Suction Lift Too High

too much friction loss on suction side

Head Increases When Flow Increases

impeller in backwards or pump turning in wrong direction

Pump Works For a While - Then Quits

incomplete priming
air leaks in stuffing box or suction piping
suction lift too high
entrained air
stuffing box defects - water seal piping plugged

Pump Draws Too Much Horsepower

speed too high
mechanical problems - bent or misaligned shaft; tight packing; bad wear rings; faulty
    bearings; deformed casing
head lower than rated - pumping too much liquid
siphoning - if pump discharge is lower than pump, the unit may be pumping and also
    siphoning, yielding a greater discharge than intended
specific gravity or viscosity of liquid other than that which pump was designed for

Motor Speed Too Low

is motor receiving full voltage?

Pump Vibrates - Excessive Noise

air in pump - any reason
clogging - pump; piping; foot valve
shaft misaligned - or bent
foundations not rigid
worn bearings
impeller worn or damaged
rotating part rubbing on stationary part

Pump Overheats & Seizes

no prime
air in pump
operating at too low capacity
shaft misaligned - bearings worn - parts rubbing

SOURCES CONSULTED

TEXTS AND MANUALS

American Water Works Association. Basic Science Concepts and Applications. Denver: American Water Works Association, 1980.

American Water Works Association. Hydraulics and Water Loss Control. Denver: American Water Works Association, 1984.

American Water Works Association. Sizing Water Service Lines and Meters. Denver: American Water Works Association, 1975.

American Water Works Association. Spillway Design and Practice. New York: American Water Works Association, 1966.

American Water Works Association. Water Meters: Selection, Installation, Testing, Maintenance. Denver: American Water Works Association, 1972.

Arasmith, E. Applied Practical Hydraulics. Albany: Linn-Benton Community College, 1978.

Binder, R. Fluid Mechanics. Englewood Cliffs: Prentice-Hall, 1943.

Brankert, K. Elementary Theoretical Fluid Mechanics. New York: John Wiley & Sons, 1964.

Brown, E. Hydraulics for Operators. Stoneham: Butterworth, 1985.

Chaudry, M. Applied Hydraulic Transients. New York: Van Nostrand Reinhold, 1987.

Cheremsinoff, N. Fluid Flow. Ann Arbor: Ann Arbor Science, 1981.

Esposito, A. Fluid Power With Applications. Englewood cliffs: Prentice-Hall, 1980.

French, R. Open Channel Hydraulics. New York: McGraw-Hill, 1985.

Giles, R. Fluid Mechanics and Hydraulics. New York: McGraw-Hill, 1962.

Hammer, J. Hydrology and Quality of Water Resources. New York: John Wiley & Sons, 1981.

Hunsacker, J. and Rightmire, B. Engineering Applications of Fluid Mechanics. New York: McGraw-Hill, 1947.

Karassik, I. Centrifugal Pump Clinic. New York: Marcel Dekker, 1981.

Kanen, J. Applied Hydraulics. New York: Holt, Rinehart & Winston, 1986.

King, H., Wisler, C., and Woodburn, J. Hydraulics. New York: John Wiley & Sons, 1922.

King, H. Handbook of Hydraulics. New York: McGraw-Hill, 1954.

Lambeck, R. Hydraulic Pumps and Motors. New York: Marcel Dekker, 1983.

Lewis Publishers. <u>Submersible Sewage Pumping Systems Handbook</u>. Chelsea: Lewis Publishers, 1985.

Linsley, R. and Franzini, J. <u>Water Resources Engineering</u>. New York: McGraw-Hill, 1964.

McNickle, L. <u>Simplified Hydraulics</u>. New York: McGraw-Hill, 1966.

Parr, B. <u>Treatment Plant Hydraulics for Environmental Engineers</u>. Englewood Cliffs: Prentice-Hall, 1984.

Prasuhn, A. <u>Fundamentals of Hydraulic Engineering</u>. New York: Holt, Rinehart & Winston, 1938.

Smith, P. <u>BASIC Hydraulics</u>. London: Butterworth, 1982.

Stewart, H. <u>Pumps</u>. Boston: T. Audel & Co., 1970.

Valentine, H.R. <u>Water in the Service of Man</u>. Middlesex: Penguin Books, 1967.

Van Haveren, B. <u>Water Resource Measurements</u>. Denver: American Water Works Association, 1986.

Vesilind, P., and Peirce, J. <u>Environmental Engineering</u>. Boston: Butterworth, 1982.

Viessman, W., and Hammer, M. <u>Water Supply and Pollution Control</u>. New York: Harper & Row, 1985.

Walker, R. <u>Pump Selection</u>. Ann Arbor: Ann Arbor Science, 1972.

Walski, T. <u>Analysis of Water Distribution Systems</u>. Berkshire: Van Nostrand Reinhold, 1984.

Water Pollution Control Federation. <u>Design of Wastewater and Stormwater Pumping Stations</u>. Washington DC: Water Pollution Control Federation, 1981.

PERIODICALS

"Accurate Flow Measurements a Necessity". <u>Opflow</u>, February 1985.

"An Engineering Approach To Waterline Thrust Restraint". <u>Public Works</u>, August 1989.

"Causes and Control of Water Hammer". <u>Opflow</u>, September 1983.

"Construction and Use of a Pitot Gage". <u>Opflow</u>, February 1989.

"Developing System Head Curve for Water Distribution Pumping". <u>Journal, American Water Works Association</u>, July 1989.

French, William. "Understanding and Choosing a Centrifugal Pump". <u>Water Engineering and Management</u>, August 1987.

Gierer, Bill. "Pumps - Mechanical Hearts of the Water Industry". <u>Opflow</u>, April 1982.

Hilsdon, Charles. "Using Pump Curves and Valving Techniques for Efficiency Pumping". <u>Journal, American Water Works Association</u>, March 1982.

Jones, Larry S. "Thrust in Underground Pipe". <u>Water Engineering & Management</u>, November 1988.

"Improved Meter Accuracy Reduces District's Problems". <u>Public Works</u>, July 1989.

Holstrom, John. "Another Look at the Versatile Venturi". <u>Water Engineering &</u>
    <u>Management</u>, November 1987.

Kroon, J. and Stoner, M. and Junt, W. "Water Hammer: Causes and Effects".
    <u>Journal, American Water Works Association</u>, November 1984.

"Purposes and Limits of Pressure Reducing Valves". <u>Opflow</u>, January 1988.

Reid, Anthony. "Selecting Pipelines to Achieve Effective Energy Conservation". <u>Water</u>
    <u>and Sewage Works</u>, November 1980.

Seruga, Ed. "Sizing and Selecting Modern Water Meters". <u>Opflow</u>, July 1986.

Sharp, W. and Walski, T. "Predicting Internal Roughness in Water Mains". <u>Journal,</u>
    <u>American Water Works Association</u>, November 1988.

Salvina, Louis. "What to Look For in Centrifugal Pump Maintenance". <u>Opflow</u>,
    February 1987.

Smith, Earl C. "Watch Your Gauges". <u>Operations Forum</u>, August 1986.

Tarrant, J. "Pipe Flow Measurement By Orifice". <u>Water and Sewage Works</u>, 1966.

Walski, T., Gessler, J. and Sjostrom, J. "Selecting Optimal Pipe Sizes For Water
    Distribution Systems". <u>Journal, American Water Works Association</u>, February
    1988.

Whitworth, Robert. "Introduction to Manometers". <u>Digester/Over the Spillway</u>,
    September 1987.

INDEX